ELECTROANALYSIS OF BIOLOGICALLY IMPORTANT COMPOUNDS

ELLIS HORWOOD SERIES IN ANALYTICAL CHEMISTRY

Series Editors: Dr MARY MASSON, University of Aberdeen,
and Dr JULIAN F. TYSON, Amherst, USA
Consultant Editors: Prof. J. N. MILLER, Loughborough University of Technology, and
Dr R. A. CHALMERS, University of Aberdeen

S. Alegret **Developments in Solvent Extraction**
S. Allenmark **Chromatographic Enantioseparation—Methods and Applications**
G.E. Baiulescu, P. Dumitrescu & P.Gh. Zugravescu **Sampling**
H. Barańska, A. Łabudzińska & J. Terpiński **Laser Raman Spectrometry**
G.I. Bekov & V.S. Letokhov **Laser Resonant Photoionization Spectroscopy for Trace Analysis**
K. Beyermann **Organic Trace Analysis**
O. Budevsky **Foundations of Chemical Analysis**
J. Buffle **Complexation Reactions in Aquatic Systems: An Analytical Approach**
D.T. Burns, A. Townshend & A.G. Catchpole **Inorganic Reaction Chemistry: Vol. 1: Systematic Chemical Separation**
D.T. Burns, A. Townshend & A.H. Carter **Inorganic Reaction Chemistry: Vol. 2: Reactions of the Elements and their Compounds: Part A: Alkali Metals to Nitrogen, Part B: Osmium to Zirconium**
J. Churáček **New Trends in the Theory & Instrumentation of Selected Analytical Methods**
E. Constantin & A. Schnell **Mass Spectrometry**
R. Czoch & A. Francik **Instrumental Effects in Homodyne Electron Paramagnetic Resonance Spectrometers**
T.E. Edmonds **Interfacing Analytical Instrumentation with Microcomputers**
Z. Galus **Fundamentals of Electrochemical Analysis, Second Edition**
S. Görög **Steroid Analysis in the Pharmaceutical Industry**
T. S. Harrison **Handbook of Analytical Control of Iron and Steel Production**
J.P. Hart **Electroanalysis of Biologically Important Compounds**
T.F. Hartley **Computerized Quality Control: Programs for the Analytical Laboratory**
Saad S.M. Hassan **Organic Analysis using Atomic Absorption Spectrometry**
M.H. Ho **Analytical Methods in Forensic Chemistry**
Z. Holzbecher, L. Diviš, M. Král, L. Šůcha & F. Vláčil **Handbook of Organic Reagents in Inorganic Chemistry**
A. Hulanicki **Reactions of Acids and Bases in Analytical Chemistry**
David Huskins **Electrical and Magnetic Methods in On-line Process Analysis**
David Huskins **Optical Methods in On-line Process Analysis**
David Huskins **Quality Measuring Instruments in On-line Process Analysis**
J. Inczédy **Analytical Applications of Complex Equilibria**
M. Kaljurand & E. Küllik **Computerized Multiple Input Chromatography**
S. Kotrlý & L. Šůcha **Handbook of Chemical Equilibria in Analytical Chemistry**
J. Kragten **Atlas of Metal-Ligand Equilibria in Aqueous Solution**
A.M. Krstulović **Quantitative Analysis of Catecholamines and Related Compounds**
F.J. Krug & E.A.G. Zagotto **Flow Injection Analysis in Agriculture and Environmental Science**
V. Linek, V. Vacek, J. Sinkule & P. Beneš **Measurement of Oxygen by Membrane-Covered Probes**
C. Liteanu, E. Hopirtean & R. A. Chalmers **Titrimetric Analytical Chemistry**
C. Liteanu & I. Rîcă **Statistical Theory and Methodology of Trace Analysis**
Z. Marczenko **Separation and Spectrophotometric Determination of Elements**
M. Meloun, J. Havel & E. Högfeldt **Computation of Solution Equilibria**
M. Meloun, J. Militky & M. Forina **Chemometrics in Instrumental Analysis: Solved Problems for the IBM PC**
O. Mikeš **Laboratory Handbook of Chromatographic and Allied Methods**
J.C. Miller & J.N. Miller **Statistics for Analytical Chemistry, Second Edition**
J.N. Miller **Fluorescence Spectroscopy**
J.N. Miller **Modern Analytical Chemistry**
J. Minczewski, J. Chwastowska & R. Dybczyński **Separation and Preconcentration Methods in Inorganic Trace Analysis**
D. Pérez-Bendito & M. Silva **Kinetic Methods in Analytical Chemistry**
B. Ravindranath **Principles and Practice of Chromotography**
V. Sedivеč & J. Flek **Handbook of Analysis of Organic Solvents**
O. Shpigun & Yu. A. Zolotov **Ion Chromatography in Water Analysis**
R.M. Smith **Derivatization for High Pressure Liquid Chromatography**
R.V. Smith **Handbook of Biopharmaceutic Analysis**
K.R. Spurny **Physical and Chemical Characterization of Individual Airborne Particles**
K. Štulík & V. Pacáková **Electroanalytical Measurements in Flowing Liquids**
J. Tölgyessy & E.H. Klehr **Nuclear Environmental Chemical Analysis**
J. Tölgyessy & M. Kyrš **Radioanalytical Chemistry, Volumes I & II**
J. Urbanski, *et al.* **Handbook of Analysis of Synthetic Polymers and Plastics**
M. Valcárcel & M.D. Luque de Castro **Flow-Injection Analysis: Principles and Applications**
C. Vandecasteele **Activation Analysis with Charged Particles**
F. Vydra, K. Štulík & E. Juláková **Electrochemical Stripping Analysis**
N. G. West **Practical Environmental Analysis using X-ray Fluorescence Spectrometry**
J. Zupan **Computer-supported Spectroscopic Databases**
J. Zýka **Instrumentation in Analytical Chemistry**

ELECTROANALYSIS OF BIOLOGICALLY IMPORTANT COMPOUNDS

J. P. HART
Department of Science
Bristol Polytechnic

ELLIS HORWOOD
NEW YORK LONDON TORONTO SYDNEY TOKYO SINGAPORE

First published in 1990 by
ELLIS HORWOOD LIMITED
Market Cross House, Cooper Street,
Chichester, West Sussex, PO19 1EB, England

A division of
Simon & Schuster International Group
A Paramount Communications Company

Typeset in Times by Ellis Horwood Limited
Printed and bound in Great Britain
by Bookcraft (Bath) Limited, Midsomer Norton, Avon

British Library Cataloguing in Publication Data

Hart, J. P.
Electroanalysis of biologically important compounds.
1. Biochemistry. Chemical analysis
1. Title
574.19285
ISBN 0-13-252107-5

Library of Congress Cataloging-in-Publication Data

Hart, J. P.
Electroanalysis of biologically important compounds /
J. P. Hart.
p. cm. — (Ellis Horwood series in analytical chemistry)
ISBN 0-13-252107-5
1. Electrochemical analysis. 2. Biomolecules — Analysis.
I. Title. II. Series.
QP519.9.E42H37 1990
574.19'285–dc20 90-4760
CIP

Table of contents

This book is dedicated to Morag and Lauren

Preface

In recent years, the demand for highly sensitive and selective methods for the determination of biomolecules has been increasing. In part, this has resulted from advances being made in medical research. In this area analytical methods are often required for the determination of ultra-trace levels of endogenous compounds in a variety of human body fluids; this probably presents the greatest analytical challenge. Studies of a nutritional nature, such as the determination of the vitamin requirements of the elderly, often require similar analytical methods. Growing interest in measuring low levels of biomolecules has also resulted from developments in biotechnology. One aspect is the requirement for careful monitoring of nutrients during manufacturing processes. Reliable and economic methods are also required for the analysis of biologically important compounds in the quality control of food and pharmaceutical products. In nucleic acid research, interest is being shown in studying, *in vitro*, the effects of certain chemical agents and radiation on nucleic acids; the ability of analytical techniques to provide information concerning these effects may, in future, eliminate the need to use live animals in such studies.

The object of this book is to demonstrate that modern electroanalytical techniques can be used for determinations of the type outlined above, as well as other related applications. A description of the electrochemical behaviour of selected biomolecules has been given, together with an explanation of the ways in which this may be exploited for their quantitative analysis. It is hoped that this approach will assist and encourage readers to develop similar strategies for other analytes. In some cases, electrochemical methods have been discussed in detail and a description of sample pretreatment, as well as other important methodological steps, have been included. The intention is that workers could, if desired, ascertain fairly quickly the potential of such methods before consulting the original source, although of course the full paper should be consulted for a thorough understanding of the procedure. Often, a selection of methods has been presented for the determination of a specific compound, or group of compounds, in order to illustrate the various strategies that may be adopted. Therefore, it is hoped that this book will both provide a basic understanding of the electrochemical processes that occur for certain classes of biomolecules, and also serve as a source of electroanalytical methods for their analysis.

It should be mentioned at the outset that only techniques involving current measurements have been described; these include modern polarographic and voltammetric techniques, amperometry, and electrochemical detection following liquid chromatography and flow injection analysis. These techniques are described in Chapter 1, together with basic theory and essential equipment and methodology. The remaining four chapters are concerned with the application of these techniques to the determination of selected classes of biomolecules, which the author considers are of great current interest. Chapter 2 is concerned with the electroanalysis of purine and pyrimidine derivatives, including nucleic acids; this is followed by a chapter on amino acids, peptides and proteins. Chapter 4 describes electrochemical methods for the determination of fat-soluble and water-soluble vitamins. The final chapter discusses the electroanalysis of several classes of coenzymes. In these applications chapter, examples involving electrochemical sensors, and biosensors, have been included; in some cases, the appropriate device and its construction have been described together with the application.

I would like to thank the authors and publishers who kindly permitted me to reprint material originally published by them; I am particularly grateful to those authors who sent me their original figures and saved me considerable effort. I would also like to express by gratitude to Steve Wring for carefully proof-reading the manuscript and for helpful suggestions. Finally, I would like to express my thanks to my wife, Morag, and daughter, Lauren, for their patience and understanding while this book was in preparation.

Symbols and abbreviations

A	electrode surface area
AA	ascorbic acid
a.c.	alternating current
AdSV	adsorptive stripping voltammetry
ASV	anodic stripping voltammetry
BSA	bovine serum albumin
C	concentration
CME	chemically modified electrode
CoPC	cobalt phthalocyanine
CSF	cerebrospinal fluid
CSV	cathodic stripping voltammetry
δ	Nernst diffusion layer thickness
D	diffusion coefficient
d.c.	direct current
DMAP1	*p*-*N*,*N*,-dimethylaminophenylisothiocyanate
DME	dropping mercury electrode
DNA	deoxyribonuclic acid
DOPAC	3,4-dihydroxyphenylacetic acid
DPP	differential-pulse polarography
DPV	differential-pulse voltammetry
E or U	electrode potential
$E_{1/2}$	half-wave potential
$E^{\ominus}$	standard electrode potential
E_p	peak potential
F	Faraday constant
FIA	flow injection analysis
GABA	γ-amino butyric acid
GSH	reduced glutathione
GSSG	oxidized glutathione
HCG	human chorionic gonadotrophin

HDV	hydrodynamic voltammogram
HMDE	hanging mercury drop electrode
i_d	limiting diffusion current
i_p	peak current
IECME	immobilized enzyme chemically modified electrode
HX	hypoxanthine
HXR	inosine
L-AAO	L-amino acid oxidase
LCEC	liquid chromatography with electrochemical detection
LCUV	liquid chromatography with ultraviolet detection
L-dopa	L-3,4-dihydroxyphenylalanine
LSV	linear-sweep voltammetry
MK	menaquinone
n	number of electrons
NAD/NADH	nicotinamide adenine dinucleotide
NADP/NADPH	nicotinamide adenine dinucleotide phosphate
NPP	normal-pulse polarography
OPA	orthophthalaldehyde
PGE	pyrolytic graphite electrode
PL	pyridoxal
PLP	pyridoxal 5′-phosphate
PM	pyridoxamine
PMP	pyridoxamine 5′-phosphate
PN	pyridoxine
PteGLU	pteroylglutamic acid, folic acid
RBP	retinol binding protein
SCE	saturated calomel electrode
SMDE	static mercury drop electrode
SSA	sulphosalicylic acid
T	thiamine
tp	pulse time
TCA	trichloroacetic acid
THC	tetrahydrocannabinol
TMP	thiamine monophosphate
TMT	trimethyl tin
TNBs	2,4,6-trinitrobenzene-l-sulphonic acid
TPP	thiamine pyrophosphate
U	average volume flow rate
ω	angular velocity
X	xanthine

1

Introduction

The purpose of this chapter is to introduce the electroanalytical techniques and methodology which appear in many of the examples described in the following chapters. It should be mentioned at the outset that the discussion is mainly intended for readers who are not familar with these techniques, and are perhaps about to pursue their application for the first time. For a more in-depth and thorough treatment of this subject several excellent monographs and reviews are recommended [1–7]. In addition, the recent monograph by Wang [8] is recommended to readers who may be interested in applying electroanalytical techniques in clinical chemistry and laboratory medicine.

This chapter is broadly divided into two sections: the first section deals with voltammetry and polarography and the second with electrochemical detection in liquid chromatography (LCEC) and flow-injection analysis (FIA-EC).

1.1 VOLTAMMETRY AND POLAROGRAPHY: TECHNIQUES AND METHODOLOGY

Voltammetry, as the name suggests, is a current–voltage technique; the recording of current versus potential is termed a voltammogram. Polarography is a special branch of voltammetry where the working electrode is a dropping mercury electrode (DME); in this case, the recording of current versus potential is called a polarogram. The earliest polarographic technique was developed in the 1920s by Heyrovsky and is known as classical direct current (d.c.) polarography [9,10]. This technique has now been superceded by modern variants that possess much greater sensitivity and selectivity. These techniques employ one of a variety of excitation waveforms; most are readily capable of determining micromolar concentrations. In addition, stripping voltammetry, which involves preconcentration of the analyte at the working electrode prior to the voltammetric measurement, has extended the limits of detection for some compounds to the sub-nanomolar level. In this section some important modern voltammetric techniques will be discussed, however, since it is instructive to have an understanding of d.c. polarography this will be described first; this will include some basic theory and an outline of the main limitations of the technique in quantitative analysis. Following the section on techniques are two sections concerned with some important aspects of methodology.

1.1.1 Techniques

1.1.1.1 Direct-current (d.c.) polarography

The classic d.c. polarographic technique involves the application of a slow linearly increasing potential sweep to a DME; Fig. 1.1a shows this potential–time wave form.

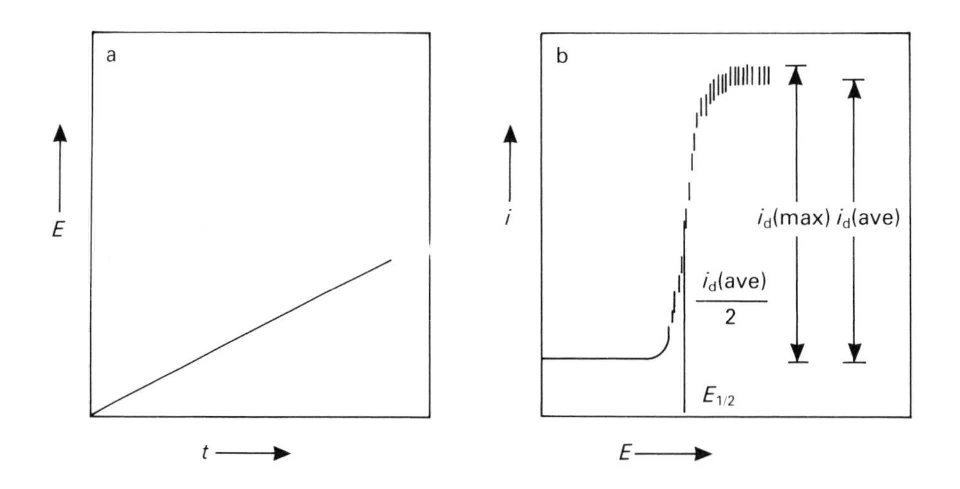

Fig. 1.1 — (a) Excitation wave form and (b) response obtained in d.c. polarography. (Adapted from [4].)

The resulting current–voltage curve is an S shape comprised of a series of oscillations that are the result of the continual growing and dislodging of the mercury drops (Fig. 1.1b). When the polarogram is the result of a reduction process of the type

$$O + ne^- \leftrightharpoons R \tag{1.1}$$

then n electrons are donated by the electrode to each molecule of the oxidized species O; therefore, the current is cathodic in nature and the reduced species R is formed at the electrode surface. When Eq. (1.1) is a strictly reversible reaction and mass transfer is diffusion-controlled, n may be determined from:

$$E = E_{1/2} + \frac{RT}{nF} \ln\left(\frac{i_d - i}{i}\right) \tag{1.2}$$

where E is the applied potential (V), R is the gas constant (8.314 J K^{-1} mol), F is the Faraday constant (96 485 C Eq.$^{-1}$), T is absolute temperature (K), i is the current on the rising part of the i–E curve, i_d is the limiting diffusion current and $E_{1/2}$ is the half-wave potential, which are measured in the manner shown in Fig. 1.1b. Equation 1.2 is known as the Heyrovsky–Ilkovic equation; at a temperature of 25°C this becomes

$$E = E_{1/2} + \frac{0.059}{n} \log\left(\frac{i_d - i}{i}\right) \tag{1.3}$$

Therefore, a plot of E versus log $(i_d - i)/i$ gives a slope of $0.059/n$, from which n can be found.

Organic electrochemical reactions usually involve the participation of protons; reduction processes are of the type

$$O + ne^- + mH^+ \rightleftharpoons R \tag{1.4}$$

where m is the number of protons reacting with each molecule of O. For a diffusion-controlled process, that occurs at 25°C, the following equation applies:

$$E_{1/2} = E^0_{1/2} - 0.059\, m/n\, pH \tag{1.5}$$

where, $E^0_{1/2}$ is the half-wave potential at a pH value of 0. Equation (1.5) shows that for organic substance undergoing reduction involving both electrons and protons $E_{1/2}$ becomes more negative with increasing pH. Thus a plot of $E_{1/2}$ versus pH can be used to determine m provided that n is known. In addition $E_{1/2}$ values may be of value for qualitative purposes but great care has to be exercised and pH as well as other conditions (e.g. buffer ionic strength) have to be known. Although only reduction processes were considered above, similar equations apply to oxidation reactions, but the current is anodic in nature. Thus useful information concerning the electrochemical characteristics of an electrode reaction can be found by d.c. polarography because the theoretical equations have been well tried and tested.

The above discussion was concerned with the application of d.c. polarography to obtain information on reversible electrochemical processes occurring at a DME. Therefore, some explanation concerning the concept of electrochemical reversibility is appropriate.

For reversible reactions of the type shown in equations 1.1 and 1.4, where O and R are electrochemically interconvertible, and where the kinetics of the reaction are very rapid, the concentrations of the oxidized (C_O) and reduced (C_R) forms at the electrode surface are given by the Nernst equation ($E = E^{\ominus} + (RT/nF)\ln(C_O/C_R)$, where $E^{\ominus}$ is the standard electrode potential of the redox couple and the other terms were described earlier). For a reversible process which occurs at a DME the $E^{\ominus}$ value is equal to the $E_{1/2}$ measured from the d.c. polarogram, provided that the solution conditions are the same. In addition, C_O and C_R can also be found from the d.c. polarogram, which leads to Eqs. (1.2) and (1.3) given above; the slope of the plot obtained from (1.3) is therefore $0.059/n$ V for a reversible reaction. Thus polarographically reversible processes can be characterized by both the $E_{1/2}$ and slope measurements obtained from d.c. polarograms. Irreversible processes can likewise be characterized from the polarograms, since the $E_{1/2}$ value does not equal the E^0 value, and the slope is smaller than $0.059/n$ V. However, it should be mentioned that electrochemical processes may show reversible characteristics using the citeria described with the d.c. technique but they may appear to be irreversible when diagnostic tests are employed using other techniques; this may occur for example when techniques employing rapid scan rates are used, e.g. linear-sweep voltammetry at a stationary mercury electrode (see sections 1.1.1.6). It is not only different techniques which affect the apparent reversibility of electrochemical reactions but the nature of electrode material. Thus a reduction process may appear to be fully reversible using a DME but when a glassy carbon electrode (see section 1.1.3) is employed it may show irreversible behaviour.

It is worth discussing at this point the concept of overpotential, or more specifically activation overpotential (η_a) since this will be mentioned in later sections. It has been mentioned above that for a reversible process $E_{1/2} = E^{\ominus}$; when the process is not truly reversible the $E_{1/2}$ value for a reduction process may be shifted several hundred millivolts more negative that the $E^{\ominus}$ (oxidation processes may be shifted to more positive overpotentials). The difference between these values is called the overpotential i.e. $\eta_a = E_{1/2} - E^{\ominus}$; the more irreversible the reaction the larger is the overpotential. Methods of minimizing this type of overpotential are particularly important in the field of electrochemical sensor development (see section 1.1.3). Large η_a values might be expected to reduce selectivity or even shift the required applied potential to unusable values. Thus electrode materials which show low values of η_a for the analyte under investigation, are highly desirable.

One further concept worth mentioning here is quasi-reversibility, which is commonly encountered when using cyclic voltammetry (see section 1.1.1.6). In this case, an electrochemical reaction of the type shown in (1.1) may occur at a particular electrode; however, the separation between the resulting anodic and cathodic voltammetric peaks is much higher then that predicted for a versible process. For a more detailed explanation of the above concepts several of the references mentioned earlier are recommended [1–3].

As mentioned above, the use of d.c. polarography in quantitative analysis is no longer commonplace but the principles involved are still worthy of mention. The parameter of interest in quantitative analysis is the limiting diffusion current (i_d); this is related to concentration as described by the Ilkovic equation:

$$i_d = knD^{1/2}m^{2/3}t^{1/6}C \tag{1.6}$$

where, i_d is in microamps, D is the diffusion coefficient (cm^2 s^{-1}), m is the rate of flow of mercury from the electrode (mg s^{-1}), t is the drop time (s), k is a constant (607 when taking mean of oscillations and 708 when taking maximum oscillations), C is bulk concentration of electroactive species (mmol dm^{-3}). In practice, all parameters except concentration are kept constant and Eq. (1.6) reduces to

$$i_d = KC \tag{1.7}$$

therefore, i_d is directly proportional to concentration.

Concentrations of electroactive species may be determined down to only about 5×10^{-5} M using d.c. polarography. The limitation in sensitivity arises mainly from interference from capacity current (i_c), which is due to charging of a double layer of ions around the electrodes. Other detrimental factors arise from the continuous application of the potential and measurement of current during the growth of the mercury drop; this results in the rather awkward $i-E$ profile (Fig. 1.1b) and also causes an extended diffusion layer to form with consequent reduction in flux (this is defined as the number of molecules penetrating a unit area of an imaginary plane in a unit of time ($mol\,cm^{-2}\,s^{-1}$)) [11]. Modern polarographic and voltammetric techniques have generally sought to address these shortcomings of d.c. polarography; some of the more important techniques are described in the following sections.

1.1.1.2 Alternating-current (a.c.) polarography

The a.c. polarographic technique involves the application of a potential wave form of the type shown in Fig. 1.2a to a DME. This may be envisaged as a linearly increasing

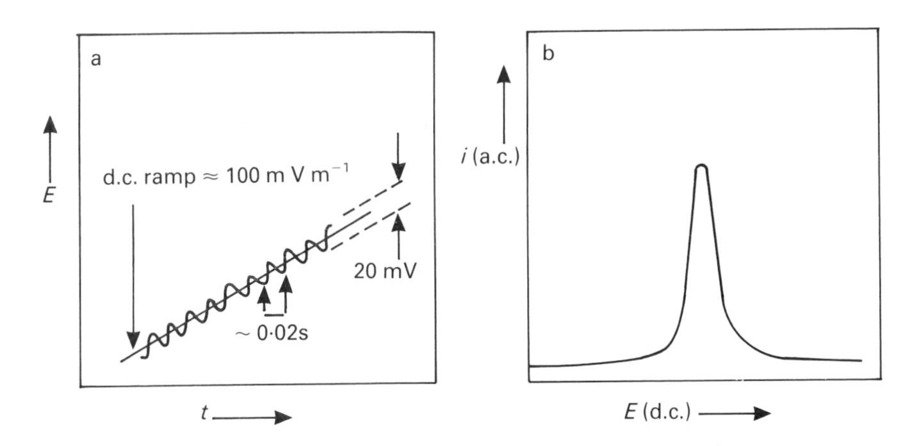

Fig. 1.2 — (a) Excitation wave form and (b) response obtained in a.c. polarography. (Adapted from [4].)

d.c. potential ramp upon which is superimposed a small alternating voltage with an amplitude of about 10–20 mV. The current flowing through the cell will contain both a.c. and d.c. components, but when the former is plotted as a function of the applied d.c. potential a symmetrical peak shape is obtained (Fig. 1.2b).

A.c. polarography is more sensitive then d.c. polarography only because it is possible to use electronic techniques to discriminate between faradaic and capacity current. This is achieved through the use of phase-sensitive detectors, which are based on the differences in phase angle between faradaic and capacity current. The former shows a phase angle of 45° for a diffusion-controlled reaction, while the latter shows a phase angle of 90° relative to the applied sinusoidal potential. Therefore, the detector operates by effectively rejecting the 90° component of the current and measuring only faradaic current [4].

For a completely reversible electrochemical process the peak potential corresponds to the $E_{1/2}$ value; however, for irreversible electrode reactions this is no longer true. In addition, the faradaic current that results from an irreversible process is smaller than that obtained for a reversible process in which the number of electrons transferred is the same. This difference arises, in a reduction process, because negligible reoxidation of the reduced species occurs during the application of the wave form. The detection limit for a species undergoing a reversible electrochemical reaction is about 10^{-6} M.

1.1.1.3 Pulse polarography/voltammetry

Pulse polarographic techniques were developed by Barker and Gardner [12] in order to improve the polarographic performance and to lower the detection limits for electroactive species. As mentioned above, the capacity current is the major source of interference in polarography (and other forms of voltammetry), and the pulse wave forms were designed to discriminate against this and measure the required faradaic current. This is achieved by applying a potential pulse, with a short duration, near to the end of the mercury drop life; this is then repeated at exactly the same time in the life of the following drops. Since capacity currents decrease more rapidly to zero than faradaic current it is possible to discriminate against the former by making a current measurement at the end of a fixed pulse period. There are basically two variations of this technique, namely: normal pulse polarography (NPP) and differential pulse polarography (DPP).

The applied wave form employed in NPP is shown in Fig. 1.3a; a pulse is applied

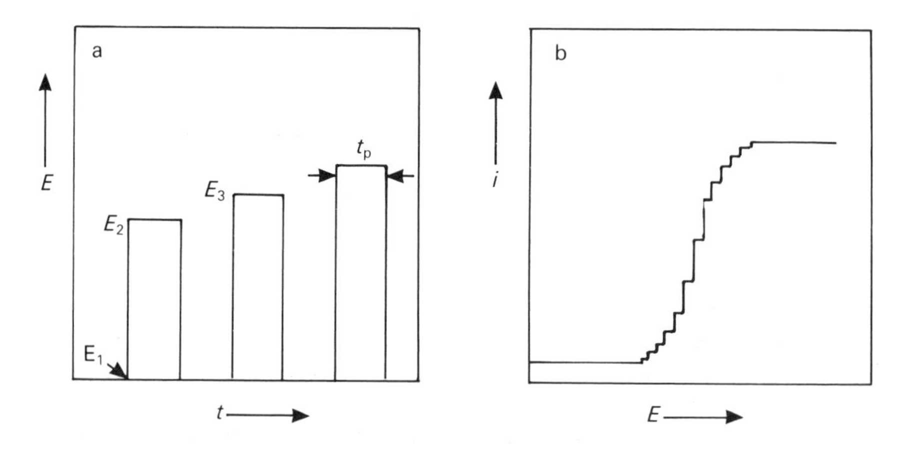

Fig. 1.3 — (a) Excitation wave form and (b) response obtained in normal-pulse polarography/voltammetry. (Adapted from [4].)

by stepping from an initial potential value E_1 to another potential E_2 and then returning to E_1 at a fixed time in the drop life. This process is continued by applying further pulses from E_1; the amplitude of subsequent pulses is increased by a constant amount to produce a linear potential scan. The pulse width (t_P) is normally set at about 50 ms; increases in pulse amplitude are a function of the scan rate and drop time and are typically around 5 mV. The current is sampled for a period of 15–20 ms at the end of each pulse, and a voltage analogue of the current is stored in the memory until the next pulse is applied. The response obtained is an S shape comprised of a series of steps as shown in Fig. 1.3b.

The limiting current obtained using NPP is described by the Cottrell equation [6,13]:

$$i_{NP,L} = \frac{nFAD^{1/2}C}{(\pi t_P)^{1/2}} \tag{1.8}$$

where A is the electrode area (cm^2), t_P is the pulse time and the other symbols have the same meaning as given earlier. The limit of detection for NPP is about 10^{-6}–10^{-7} M.

In DPP the wave form shown in Fig. 1.4a is applied to a DME. This may be

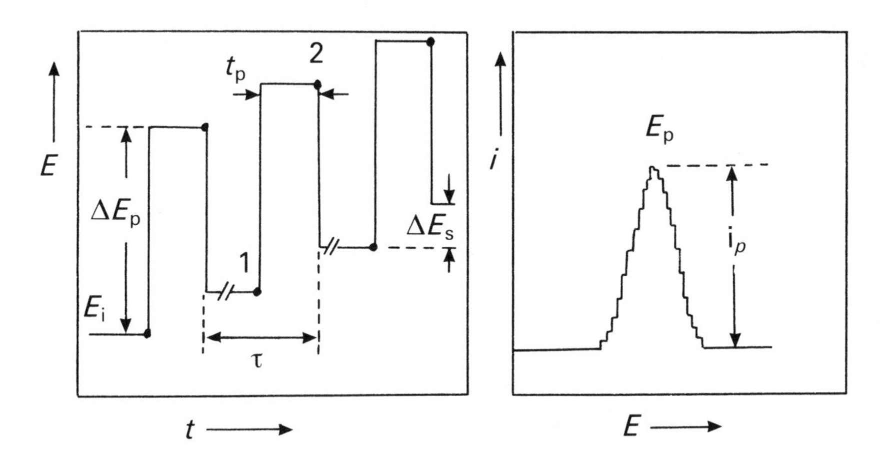

Fig. 1.4 — (a) Excitation wave form and (b) response obtained in differential-pulse polarography/voltammetry. E_i is the initial potential. (Adapted from [6].)

envisaged as the sum of a staircase (step height ΔE_s) with a pulse train [14]. As explained above, application of these pulses allows discrimination of the unwanted capacity current from the required faradaic current. However, in DPP the current is also measured just before the pulse is applied so that a current difference ($i_2 - i_1$; Fig. 1.4b) is presented as the output to the recorder. This results in a peak shape response as shown in Fig. 1.4b. Pulse amplitudes (ΔE_P) of 5–100 mV are usually employed in conjunction with drop times τ of 0.5–5 s; the pulse width, t_P, is usually set at about 50 ms, and scan rates are usually 1–5 mV s^{-1}.

The peak current in DPP is described by Eq. (1.9) when the pulse amplitude is less than RT/nF [13]:

$$i_P = \frac{n^2F^2AC}{4RT}\Delta E_P(D/\pi t_P)^{1/2} \tag{1.9}$$

Since i_P is directly proportional to n^2, the currents obtained for 2e and 3e processes would be 4 times and 9 times greater than that obtained for a 1e process when all other parameters are equal; this has important implications in quantitative analysis. In addition, i_P is also directly proportional to ΔE_P over a narrow range; however, care has to be taken when large values are employed (above 100 mV) because peak broadening occurs and this can result in loss of resolution [14,15]. The limit of detection for DPP is about 10^{-7} 10^{-8} M.

Both normal-pulse (Fig. 1.3a) and differential-pulse (Fig. 1.4a) wave forms can be applied to stationary mercury and solid electrodes; the techniques are then referred to as normal-pulse and differential-pulse voltammetry. The instrumental conditions employed in these techniques are usually the same as in the corresponding polarographic techniques and the resulting voltammograms exhibit the same basic features. The limits of detection in pulse voltammetry are often the same as in pulse polarography; however, in some cases spontaneous adsorption has been found to occur at the electrode surface, which results in greater signals for the voltammetric techniques. This phenomenon forms the basis of another electrochemical technique, known as adsorptive stripping voltammetry, which is discussed later.

1.1.1.4 Square-wave voltammetry [6]

In square-wave voltammetry a potential wave form of the type shown in Fig. 1.5a is

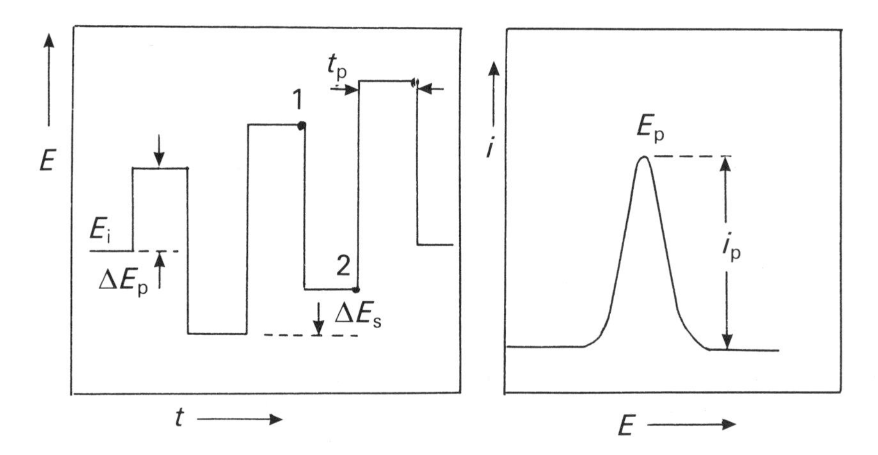

Fig. 1.5 — (a) Excitation wave form and (b) response obtained in square-wave voltammetry, E_i is the initial potential. (Adapted from [6].)

applied to the working electrode. This is esentially a staircase, with step height ΔE_S, added to which is a symmetrical pulse train with a total amplitude of $2\Delta E_P$; the period of this is τ. The output current is sampled at the end of the forward and reverse half-cycles indicated by the positions 1 and 2 respectively in Fig. 1.5a; a difference current is measured by substracting the current at 2 from that at 1. The resulting output signal is shown in Fig. 1.5b. The value of ΔE_S is about $10/n$ mV, ΔE_P is about $50/n$ mV and t_P is approximately 8 ms; therefore, the technique is somewhat more rapid than the other techniques mentioned and a complete voltammogram may be obtained in about 1 s [6]. In addition, square-wave voltammetry is also more sensitive even than DPP; this is because both forward and reverse currents are measured in the former but only the forward currents are measured in the latter.

1.1.1.5 Staircase voltammetry

This is also regarded as a pulse technique [6] and involves an excitation wave form of the type shown in Fig. 1.6a; the step heights are typically 10 mV or less and the width

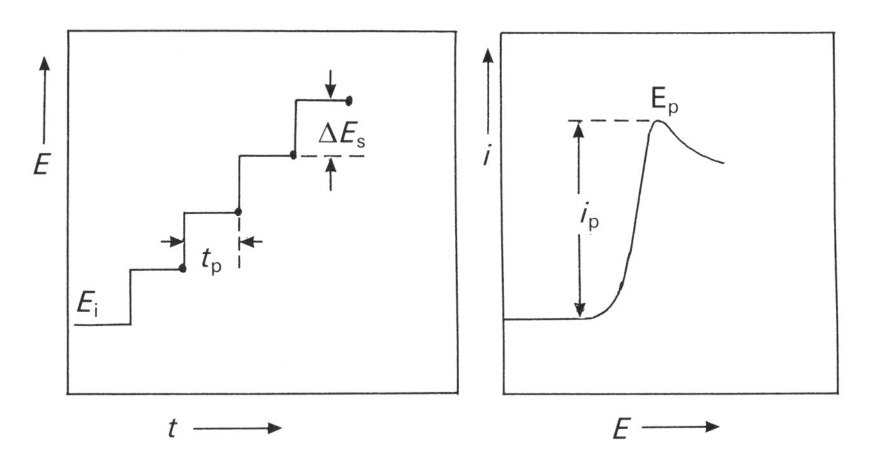

Fig. 1.6 — (a) Excitation wave form and (b) response obtained in staircase voltammetry, E_1 is the initial potential. (Adapted from [6].)

is typically 17 ms. The current is measured at the end of the step, where the capacity current has decayed appreciably but the faradaic current is still significant. The resulting output signal resembles a linear sweep voltammogram (Fig. 1.6b), which is described below. The advantage of staircase over linear sweep voltammetry is that there is a much reduced capacity current component in the signal.

1.1.1.6 Linear-sweep and cyclic voltammetry

Liner-sweep voltammetry involves the application of a rapid linear potential sweep, usually in the range 10–1000 mV s^{-1}; the resulting current voltage curve is the same as that shown in Fig. 1.6b. The peak current obtained at a planar electrode for a reversible process is described by the Randles–Sevcik equation:

$$i_P = 2.69 \times 10^5 n^{3/2} A D^{1/2} \upsilon^{1/2} C \tag{1.10}$$

where υ is the scan rate (V s^{-1}), and the other symbols have the same meaning as given earlier. It is apparent from Eq. (1.10) that i_P is proportional to scan rate and large signals are obtained with very fast scans; the limit of detection of this technique is about 10^{-6} M. Linear-sweep voltammetry is a very useful technique at most solid electrodes because rapid analysis times can be achieved [16]. However, one of the main drawbacks is the difficulty in measuring the current for a peak which occurs on the decreasing tail of a preceding peak.

In cyclic voltammetry a triangular potential waveform is applied to the working electrode as shown in Fig. 1.7a. This is achieved by rapidly scanning the potential

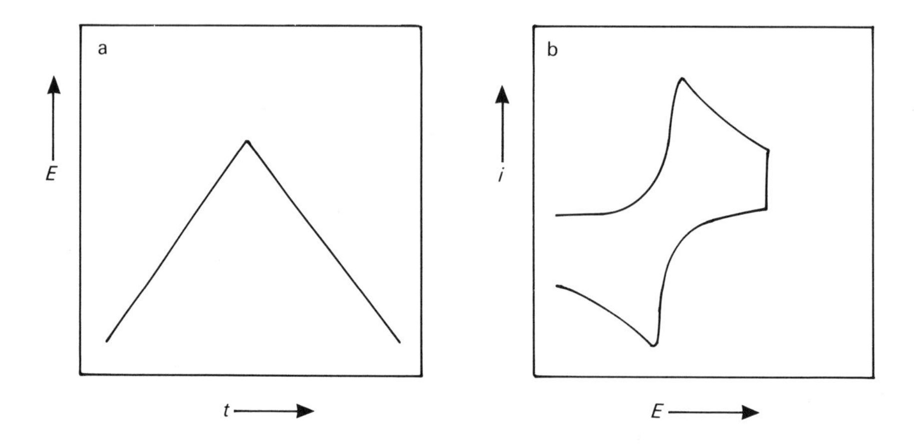

Fig. 1.7 — (a) Excitation wave form and (b) response obtained in cyclic voltammetry. (Adapted from [4].)

range of interest in a forward direction and then reversing the direction of the scan. When the electrode reaction is reversible the cyclic voltammogram exhibits peaks on both the forward and reverse scans (Fig. 1.7b). In this case, the separation between the peak potential of the anodic peak (E_{pa}) and that of the cathodic peak (E_{pc}) is $0.059/n$ V (i.e. $\Delta E_p = E_{pa} - E_{pc} = 0.059/n$) [2]. For a quasi-reversible process the ΔE_p value is greater than $0.059/n$ V. If the reaction is totally irreversible, only the peak on the forward scan is observed. Therefore this technique gives a very rapid indication of the reversibility or otherwise of electrochemical processes. It is possible in some cases to ascertain whether chemical processes follow on from the initial electrochemical process, which in turn is followed by a further electrochemical transfer (ECE mechanism); examples of this type have been included in the following chapters, Scan rates of about 10–100 mV s^{-1}, (and sometimes greater) are usually employed in this technique.

1.1.1.7 Voltammetry and amperometry in stirred solution and with rotating disc electrodes

The hydrodynamic flow of a solution over the electrode surface results in an increase in sensitivity and several electroanalytical techniques employ this particular approach. All forms of hydrodynamic voltammetry are steady-state methods in that the current at a given potential is independent of the scan direction and time [17].

In one method the electrode is stationary and the solution is stirred, usually with a magnetic stirrer. The resulting voltammograms have an S-shaped appearance, with a subsequent plateau region from which the limiting current (i_L) can be measured; this is done in a manner similar to d.c. polarography (Fig. 1.1b). The equation for the limiting current is

$$i_L = \frac{nFADC}{\delta} \tag{1.11}$$

where δ is the Nernst diffusion layer thickness; this parameter is considered to be the thickness of a stationary layer of solution immediately adjacent to the electrode surface.

For quantitative determinations it is possible to simply fix the applied potential at a position on the plateau of the wave and make current measurements at this potential; this is then considered to be amperometry in stirred solution and has the advantage that very simple instrumentation is required. In practice, one first introduces the supporting electrolyte into the voltammetric cell, applies a voltage and records the current–time profile (amperogram). Next a small volume of the sample is introduced into the cell, which results in the appearance of a step; the height of this step is usually proportional to concentration when all other parameters are kept constant. In order to introduce some degree of selectivity such methods may also incorporate enzymes. The present author and colleagues have found this method useful in conjunction with chemically modified electrodes that effectively lower the potential required for oxidation or reduction; this also improves the selectivity and often the sensitivity of the method; detection limits may be down to about 10^{-8} M under favourable conditions [18]. Bearing in mind that the current trend for electrochemical sensors, and biosensors, is that they should be simple devices [19], this approach is probably the simplest method of evaluating electrode materials that may be incorporated into such devices. Several examples of the application of amperometry in stirred solutions are included in Chapters 3 and 4.

In another approach utilizing hydrodynamically voltammetry, the electrode is rotated and the solution is unstirred; it has been considered that the hydrodynamic properties of the solution are easier to control than with stirred solutions [20,21]. The limiting current in this technique is given by [22,23]:

$$i_L = 0.620nFACD^{2/3}\upsilon^{-1/6}\omega^{1/2} \tag{1.12}$$

where ω is the angular velocity of the disc ($\omega = 2\pi N$, N is the number of revolutions per second), υ is the kinematic viscosity of the fluid ($cm^2\,s^{-1}$), C is the concentration ($mol\,cm^{-3}$) and i_L is in amps. This technique has found wide use in studies of electrode mechanisms but has not really found widespread use in quantitative analysis.

1.1.1.8 Stripping voltammetry

Stripping voltammetry involves two separate stages; there is initially a preconcentration step to accumulate the analyte into, or onto, the working electrode, which is then electrochemically stripped back into solution in the current measurement step. Fig. 1.8a shows a typical potential wave form applied to the working electrode.

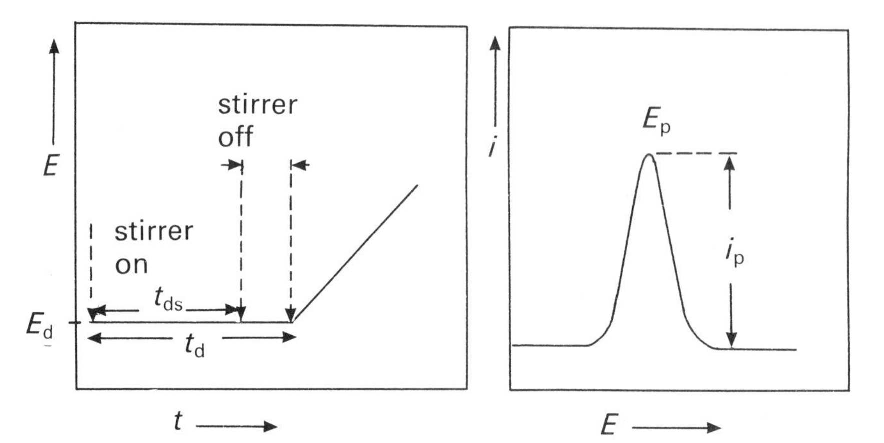

Fig. 1.8 — (a) Excitation wave form and (b) response obtained in stripping voltammetry.

During the preconcentration step, at a deposition potential E_d, the solution is usually stirred and this is discontinued before scanning (normally 30 s is allowed for the solution to become quiescent). The deposition time t_d is the total time of the preconcentration step including the quiescent period; however, for analytical purposes it is usual to investigate the effect of the deposition time with stirring, t_{ds}, while keeping the quiescent period constant. Clearly, the longer the deposition times used the greater the resulting stripping signal; the optimum t_{ds} is usually a trade-off between peak magnitude and the analysis time. The stripping step may involve the application of a linear potential sweep, differential pulse, square wave, etc., and Fig. 1.8b shows a typical output signal.

The accumulation step may involve one of several mechanisms. For metal ion determination the species is first reduced at negative potentials to form an amalgam which dissolves and preconcentrates into, for example, a hanging mercury drop electrode (HMDE). The accumulated species is then stripped back into solution by applying a positive-going potential waveform; this is known as anodic stripping voltammetry (ASV) which allows sub-nanomolar analyte levels to be measured. The principles and applications of this technique have been well documented [24]. The term ASV has been coined for this particular application but positive-going sweeps may also be used with other types of accumulation mechanisms which are described below.

Cathodic stripping voltammetry (CSV) involves the accumulation of a species at relatively positive potentials; a number of thiol-containing species have been shown to undergo oxidation reactions with the formation of an insoluble mercury–thiol species as a film around the mercury electrode, e.g. an HMDE. The adsorbed film may then be reduced using a cathodic scan which strips the material from the electrode surface. This technique also provides very low detection limits which are of the same order as ASV [25].

Recently, there has been growing interest in using stripping voltammetry for organic species that cannot be accumulated by electrolysis; the preconcentration processes have been mainly based on adsorptive accumulation. The technique involving such accumulation, followed by voltammetric measurement of the surface species, is known as adsorptive stripping voltammetry (AdSV) [26,27]. This technique has allowed many compounds possessing adsorption characteristics to be detected at levels similar to those obtained by ASV and CSV.

In order to obtain maximum sensitivity of the voltammetric signal in AdSV, optimum conditions for maximum adsorption should be employed during accumulation. The factors which may affect this include; accumulation potential and time, pH and ionic strength of supporting electrolyte, the presence of organic solvents, mass transport of analyte to the electrode [26,27]. Working electrode materials that have found application in this technique include: mercury, in the form of a HMDE, for compounds that can undergo reduction; and platinum, glassy carbon and carbon paste for oxidizable species. In the latter case, adsorptive accumulation of some compounds may also be accompanied by extraction into the electrode pasting material [28,29]. It has been possible to carry out preconcentration in the sample solution and then remove the electrode to a fresh supporting electrolyte for the voltammetric measurement; this can greatly improve selectivity of the procedure, particularly in biological fluids, and has been termed medium exchange. Voltammetric stripping of the accumulated species may be performed by scanning in the cathodic or anodic direction depending on the electrochemical properties of the adsorbed species. The voltammetric waveforms that may be employed for AdSV are also those mentioned above for conventional stripping techniques. There have been an increasing number of applications of AdSV to trace analysis of compounds of biological importance; including hormones [30], tranquilizers [31,32], antibiotics [33] and dopamine [34]. Further examples of this type of analysis will be described in greater detail in later chapters.

1.1.2 Experimental considerations

1.1.2.1 Supporting electrolytes, solvents and dissolved oxygen

Supporting electrolytes are required in solutions for voltammetric studies in order to conduct a current through the solution; the concentration of this must be much greater than the analyte (usually 100–1000 times greater) to prevent the analyte from migrating to the working electrode by electrostatic forces [4,35].

The most common supporting electrolytes used in aqueous solutions are: potassium and lithium chloride; dilute sulphuric and hydrochloric acids; sodium hydroxide; and buffers such as acetate, citrate, borate and phosphate. For non-aqueous voltammetry tetraalkylammonium salts (e.g. tetraethylammonium perchlorate) have been commonly used.

For voltammetric studies on some organic substances it may be necessary to carry out the investigation in an organic solvent, such as dimethylsulphoxide (DMSO), dimethylformamide (DMF), or acetonitrile. However, there are many instances when a mixture of aqueous buffer solutions containing an organic solvent such as methanol are advantageous. Under these conditions organic reactions involving protons as well as electrons are still pH-dependent; it is possible to use this phenomenon to shift the peak potentials of species to potentials where interference may not occur [36]. Therefore, preliminary investigations to study the effect of solvent concentration, buffer pH and ionic strength are usually performed.

Oxygen is capable of dissolving in aqueous solutions to produce millimolar concentrations; this can undergo reduction at a working electrode by the following mechanisms:

in acid solution

$$O_2 + 2e^- + 2H^+ \rightarrow H_2O_2 \text{ (first reduction)} \tag{1.13}$$

$$O_2 + 4e^- + 4H^+ \rightarrow 2H_2O \text{ (second reduction)} \tag{1.14}$$

in neutral or alkaline solution

$$O_2 + 2e^- + 2H_2O \rightarrow H_2O_2 + 2OH^- \text{ (first reduction)} \tag{1.15}$$

$$O_2 + 4e^- + 2H_2O \rightarrow 4OH^- \text{ (second reduction)} \tag{1.16}$$

These reduction processes occur in the potential regions between about -0.05 and -1.3 V versus a saturated calomel electrode (SCE); therefore, voltammetric waves due to dissolved oxygen can overlap with many other electroactive analytes. In addition, the products of these processes (H_2O_2 and OH^-) may react with the species to be determined and cause erroneous results. For these reasons it is usual to remove dissolved oxygen, prior to voltammetric (polarographic) analysis, when scanning in the cathodic direction. This is normally achieved by passing nitrogen gas through the solution containing analyte for about 10 min prior to the voltammetric measurement. Commercially available nitrogen contained in cylinders usually contains traces of oxygen so it is necessary to remove this before deaerating the sample solution; this is usually done by passing the gas through vanadous chloride solution contained in a dreschel bottle [37].

1.1.2.2 Quantitative procedures

Quantitative analysis using d.c. polarography is based on the Ilkovic equation, which was described earlier (section 1.1.1.1); as was also mentioned, all parameters except concentration are kept constant so that the current is directly proportional to concentration, i.e. $i_d = KC$. This same approach is adopted with the modern variants of this technique, such as differential pulse and square wave voltammetry, although in these cases it is the peak current (i_P) that is measured. It should be added that other techniques, such as a normal-pulse and rotating-disc voltammetry, exhibit voltammograms with the S shape and the limiting currents (i_L) are measured in the manner described for d.c. polarography. When all of the parameters, except concentration, in these modern variants are kept constant the general equation is:

$$i_P(i_L) = KC \tag{1.17}$$

where C is the bulk concentration, which may be in one of the commonly used concentration units e.g. millimolars, nanograms per cubic centimetre, etc., and i_P (i_L) is usually in microamps, or nanoamps.

The simplest method of quantitative voltammetric analysis is to measure the i_P or i_L values from the voltammogram of the analyte of unknown concentration and refer this to a standard calibration graph of i_P (i_L) versus concentration. The standards used for such a plot could simply be prepared in pure solution containing the supporting electrolyte. However, this method is not always reliable because the solution of unknown concentration may also contain substances that are absent from the standards; these substances may affect both current and potential values. This would often be the case in, for example, simple solvent extracts from body fluids. A more reliable technique is the method of standard addition. In one variation of this a voltammogram is recorded before, and after, the addition of a known volume of a standard solution of the analyte. In this way the identity of the analyte is confirmed because the voltammetric signal merely increases in magnitude and the peak potential (E_p), or $E_{1/2}$, is identical for both voltammograms. Assuming that a voltammetric peak response is obtained, the unknown concentration can then be found from

$$C_u = \frac{i_P V_S C_S}{(i_{P(1)} - i_{P(2)})V_1 + i_{P(2)} V_S} \tag{1.18}$$

where C_u is the unknown concentration, C_s is the concentration of the standard solution, $i_{P(1)}$ is the initial peak current, $i_{P(2)}$ is the peak current after the addition of standard solution, V_1 is the initial volume of sample and V_s is the volume of standard solution added. Of course with this technique it is important to establish first that there is linearity of response with concentration over the range of interest.

Another variation of this technique involves multiple standard additions of standard solution to the unknown and the voltammograms are recorded after each addition. If the additions of standard are small (e.g. 10 μl) compared with the original sample volume, V_0 (e.g. 10 cm^3) it is not usually necessary to make corrections for dilution over, for example, four additions. In order to determine the unknown concentration a plot of, for example, peak current (i_P) versus volume of standard added (V_1, V_2, etc.), is constructed (Fig. 1.9); the volume of standard determined where $i_P = 0$ is equivalent to unknown analyte (V_u); and from this the original concentration in the sample can be found. For example, if the concentration of standard is C_s, $\mu g\,cm^{-3}$,and the volumes V_u and V_o are in cm^3, then the unknown concentration, C_u, is given by:

$$C_u = \frac{C_s V_u}{V_o} \mu g\,cm^{-3} \tag{1.19}$$

This method would be expected to be more accurate than the method of single standard addition because any errors in the additions are averaged out; furthermore, the linearity of the calibration is confirmed.

Another approach has been employed by Brooks and coworkers [38–40] for drug analysis. This involves the addition of a range of standard amounts of the substance to be determined to control samples (whole blood, plasma, serum or urine) before any separation steps are carried out. These spiked samples are then treated in an identical way to the sample containing the unknown concentration. A calibration graph of i_P versus concentration is then constructed from the 'spiked' standards and the unknown concentration determined by reference to the calibration graph. This method avoids the need to compensate for losses of substance during the extraction procedure.

It is usual in voltammetric analysis to determine the recovery of the overall method in the same way as in other analytical techniques. In the case of body fluids

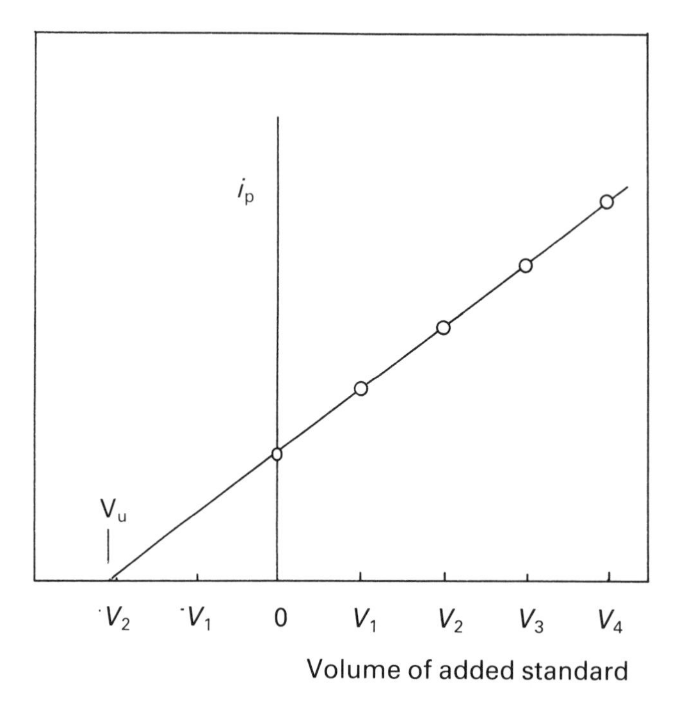

Fig. 1.9 — Multiple standard addition method of quantitation.

this may be done by spiking samples, as described above, and carrying out the analysis. In addition, a calibration graph is constructed by making standard additions directly to a control-only extract in the voltammetric cell. The recovery of the drug, or other substance, can be found by the ratio of the two calibration plots [37]. When the sample is a pharmaceutical preparation, a number of individual samples of known concentration may be analysed and the content determined by the methods of standard addition. The recovery can then be calculated by reference to the manufacturer's values. However, some caution should be exercised with this method because in some cases e.g. multivitamin preparations, a minimum value, and not the actual concentration, may be specified [41].

1.1.3 Cells, electrodes and potentiostats

Modern voltammetric analyses are usually performed in a vessel containing a three-electrode system of the type shown in Fig. 1.10. This type of cell is commercially available in a range of sizes that can accommodate 2–50 cm^3 of solution; however, smaller volumes can be investigated using a specially constructed microcell [38]. The important considerations in the design are that the electrodes should be as close as possible to one another and there should be a simple means of deaeration before analysis.

As mentioned earlier, the working electrode in polarography is the dropping mercury electrode (DME), which consists of a glass capillary, about 15 cm long and about 0.05 mm in internal diameter. The mercury electrode can also take the form of

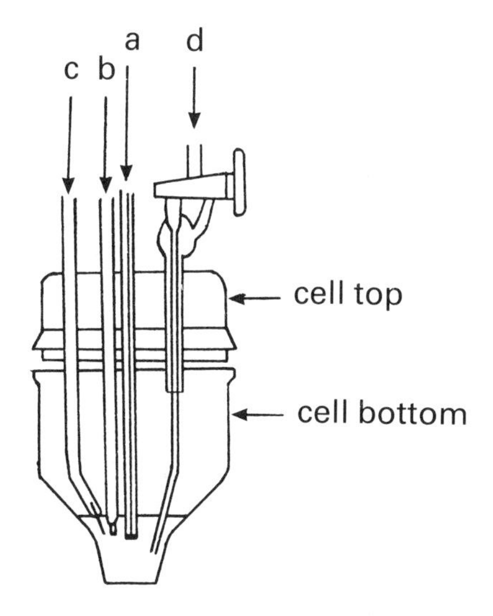

Fig. 1.10 — Voltammetric cell containing (a) working electrode, (b) reference electrode, (c) auxiliary electrode, (d) a two-way nitrogen line for deaeration. (Reprinted from [37] by courtesy of Marcel Dekker Inc.)

a hanging drop; this may be achieved by a micrometer arrangement where the mercury drop size is produced manually. However, most modern voltammetric instruments incorporate a special electrode assembly, which may be operated by gas pressure or by a solenoid; this allows the operator to select either the dropping mercury or stationary hanging drop mode. Mercury electrodes have the advantage that they can be used over wide cathodic ranges (up to about −2.0 V versus SCE); however, they are useful over only a limited anodic range due to the mercury itself becoming oxidized.

This problem has been overcome by the discovery of other electrode materials which extend the anodic range to about +2 V versus SCE. These materials include the noble metals (gold, silver and platinum), carbon paste, glassy (vitreous) carbon, and pyrolytic graphite. In recent years, there has been considerable interest in the use of chemically modified electrodes for voltammetric analysis. These electrodes have been prepared in a variety of ways, which include: by immobilization of the agent by polymerization onto carbon substrates, by mixing the agent with graphite and forming a carbon paste electrode; and by deposition onto glassy carbon and other carbon electrodes [7,42–57]. A number of chemical agents have been investigated for incorporation into these electrodes; these include metal-containing compounds in which the metal species participates in the redox processes [44,45,48,50,51,53], and organic substances which can readily undergo redox processes [46,47,49,52,54,55]. The basic concept is that suitable chemical agents act as mediators which may reduce considerably the overpotential for the reduction or oxidation of the analyte. Therefore, this approach is very useful when electrochemical reactions occur at potentials which are not accessible with the more conventional electrodes. In addition, it may also be useful to improve selectivity when using, for example amperomerty in stirred solutions [18,58].

Amperometric detection has also been employed as the basis of many biosensors [59–66]. These highly selective devices may be fabricated by immobilizing an enzyme in, or onto, the working electrode; the analyte is usually determined indirectly by one of several mechanisms. For example, glucose may be determined by measuring the hydrogen peroxide produced when glucose oxidase and oxygen participate in the oxidation reaction. The same analyte may also be determined by the use of a ferrocene-mediated reaction, in which the ferricinium ion acts as the electron carrier to the electrode instead of oxygen; this is the basis of a commercially available glucose sensor [64].

Another approach which has been used to achieve high selectivity in amperometric detection involves immunological interactions of the antibody/antigen type [19,63–64]. Since these interactions do not usually involve electroactive reactants, enzyme labelling has been employed to mediate between immunological and electrochemical processes in the biosensor. Such sensors have been fabricated by attaching antibodies to the surface of an oxygen electrode; an enzyme may be attached to a second antibody, or to an antigen (competitive configuration) [64]. The assay involves either the attachment of enzyme to, or its displacement from, the electrode in the presence of antigen. The electrode is washed to remove unbound enzyme, and then substrate is added; the resulting currents are either directly, or inversely, related to the antigen concentration. This type of sensor has been applied to the determination of a variety of substances, including insulin, albumin, immunoglobulins, and thyroxine [64].

In addition to the working electrode, modern instruments employ a reference and auxiliary electrode. The most commonly used reference electrodes are the SCE and the silver/silver chloride electrode (Ag/AgCl). The purpose of the reference electrode is to accurately monitor the applied potential 'felt' at the working electrode so that, if necessary, adjustments may be made (see below). The auxiliary electrode is normally made from platinum wire and serves as a means of applying an input potential to the working electrode.

The potentiostat is a device which can compensate for the voltage drop across a high resistance solution; if this was not done the actual potential at the working electrode would be different to the applied potential. The potentiostat contains an operational amplifier which is connected to the reference and auxiliary electrodes as shown in Fig. 1.11; the reference electrode has a constant potential and is placed as close as possible to the working electrode in the cell. The input potential wave form is applied through the auxiliary electrode to the working electrode; therefore, any voltage drop is experienced by both the reference and working electrode. In this case, the operational amplifier will apply sufficient compensating potential to the auxiliary electrode to ensure that the potential at the reference electrode tip (and therefore the working electrode) is the desired one [67].

1.2 LIQUID CHROMATOGRAPHY AND FLOW INJECTION ANALYSIS WITH ELECTROCHEMICAL DETECTION

One of the most powerful analytical techniques to emerge in recent years involves electrochemical detection following high-performance liquid chromatography (LCEC) [68–71]. This technique has been shown to possess great potential for the

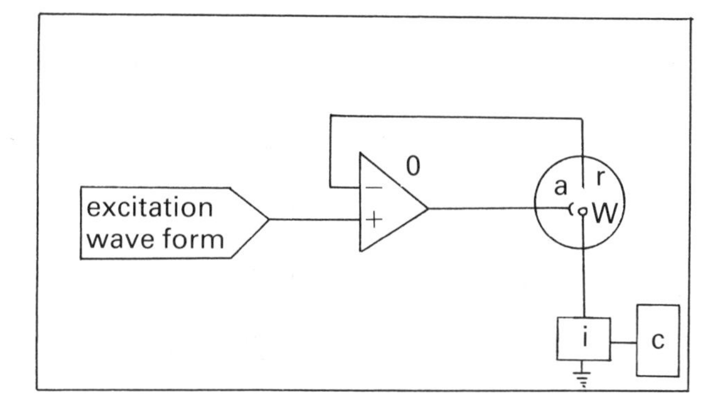

Fig. 1.11 — Three-electrode operation with potentiostatic control: (w) working electrode; (r) reference electrode; (a) auxiliary electrode; (o) operational amplifier; (i) current-to-voltage converter; (c) chart recorder.

determination of trace quantities of electroactive species of biomedical and biological interest in complex matrices [72–73]. For the most part, the selectivity of the method is achieved by the appropriate choice of the column; the vast majority of LCEC applications involve the use of reversed-phase columns. However, in some cases adequate selectivity has been achieved without a column, as in flow injection analysis (FIA); in this instance selectivity may be obtained by the choice of electrochemical detector, and/or by some form of catalytic reaction, e.g. an enzyme reaction. These FIA detectors use the same transducer technology developed for LCEC [71]. The choice of the cell and electrode configuration used for LCEC and FIA depends on the particular application; some of the more common types will be briefly mentioned here. The detectors can be broadly divided into amperometric and coulometric and will be described accordingly.

1.2.1 Amperometric detectors

In amperometric detection an electrochemical conversion occurs at the electrode surface that electrolyses much less than 100% of the analyte (typically about 1%) [74]. The amperometric detectors may be equipped with cells containing a single electrode, or with two working electrodes arranged in one of several different ways.

One of the most popular electrode arrangements, using a single working electrode, is the thin-layer cell; there are several variations of this cell type [71,75]. Fig. 1.12a shows a typical thin-layer cell design in which the working electrode is situated in one half of the cell and is parallel to the following stream of liquid. The other half of the cell contains the reference electrode and the auxiliary electrode, is made of stainless steel and forms the eluent outlet. The cell halves are separated by a gasket which may be one of several thicknesses; commercially available gaskets are typically 0.002 to 0.005 cm thick.

In a similar cell design the auxiliary electrode may be placed opposite the working electrode, with the reference electrode downstream (Fig. 1.12b) [75]. This latter design reduces the *iR* drop between the working and auxiliary electrodes; this results in greater linear ranges of calibration plots. The current produced in a thin layer cell,

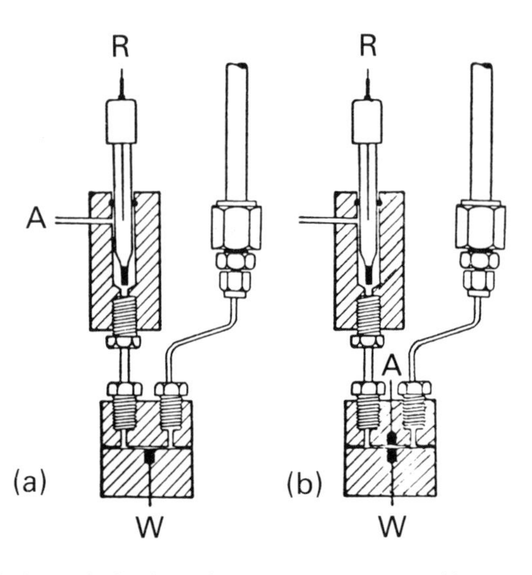

Fig. 1.12 — Thin-layer single-electrode amperometric cells, (a) auxiliary electrode (A) downstream and (b) opposite the working electrode (W), R is the reference electrode. (Reprinted from [75] by courtesy of Marcel Dekker Inc.)

with the working electrode positioned centrally in a rectangular channel, has been described by Elbicki *et al.* [76]:

$$i = 1.47nFC(DA/b)^{2/3}U^{1/3} \qquad (1.20)$$

where i is the current density (μA cm^{-2}), b is the gasket thickness (cm), U is the average volume flow rate (cm^{3}s^{-1}), C is the concentration (mM) and the other symbols have the same meaning as given earlier. From this it is apparent that the thinner the gasket the larger the current produced. This approach was used to effect an increase in current magnitude of vitamin K_1 [77]; by changing the thickness from 0.005 to 0.002 cm the current increase agreed with that predicted by Eq. (1.13). As is also apparent from this equation the current is expected to increase with flow-rate and electrode area; however, in practice these are limited because the background current also increases. As background current increases so does the level of noise and it is signal-to-noise ratio that is important in determining the sensitivity. Since the background currents are mainly due to impurities in the mobile phase it is important to use pure reagents.

Another popular cell design is the wall-jet cell which was developed by Fleet and Little [78]; in this configuration the eluent impinges normal to the electrode surface (the wall, Fig. 1.13). For some analyses, this configuration offers the advantage that it is less prone to ‘poisoning’ by the adsorption of contaminents, or the products of electrode reactions; it appears that the jet of eluent actually has a cleaning action on the electrode surface. When the cell containing the working electrode has a rectangular channel the current density is also described by Eqn. 1.20 [7,76].

Other important single working electrode detectors have been described by Stulik and Pacakova [7] and Hanekamp *et al.* [74]; these include cells containing

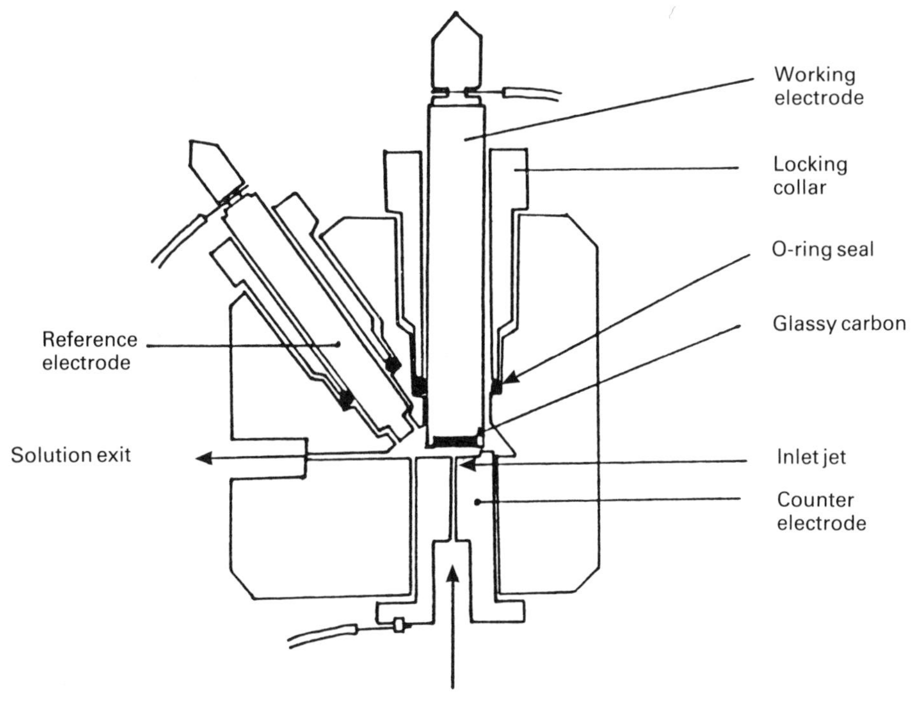

Fig. 1.13 — The wall-jet amperometric cell. (Reprinted by courtesy of EDT.)

tubular electrodes, as well as mercury electrodes, in the opposite, parallel and normal configurations.

Dual-electrode amperometric detectors are becoming increasingly important for the analysis of complex samples. The two most popular arrangements of dual electrodes in thin-layer cells are the parallel and series configurations [79–80].

In the parallel configuration, two identical working electrodes are arranged side by side so that the eluent contacts both electrodes simultaneously (Fig. 1.14a). In one mode of operation, applied potentials are selected at two points on the hydrodynamic wave, e.g. E_1 and E_2, Fig. 1.14c. The ratio of the currents i_1/i_2 is calculated and is then compared with a standard of a pure compound for confirmation of peak identity. It is also possible to operate this configuration in another mode that allows the measurement of both oxidizable and reducible species in the same injection, i.e. simultaneously. In this case, the potential of one electrode is held at a positive value that is sufficient to initiate an oxidation process (E_1; Fig. 1.14d); the other electrode is held at a potential where reduction of the analyte occurs (E_2; Fig. 1.14d).

In the series configuration, one electrode is situated upstream of the other (Fig. 1.14b). These may be operated in what is called the 'redox' mode provided that the electrochemical reaction is reversible or quasi-reversible. The upstream electrode (designated as the 'generator' electrode) is set at a potential where, for example, the

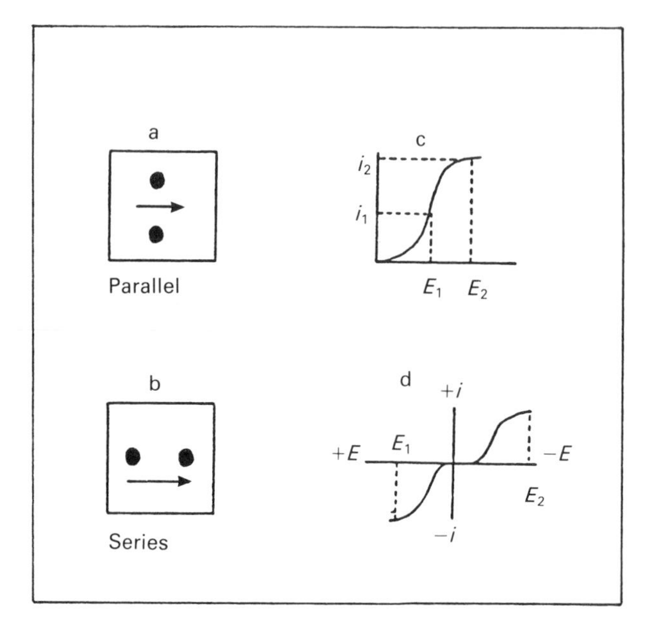

Fig. 1.14 — Dual-electrodes in thin-layer amperometric cells (for explanation of letters a–d, see text). (Reproduced from [73] by permission of the copyright holders, Elsevier Science Publishers Physical Sciences and Engineering Division.)

analyte is reduced (E_2; Fig. 1.14d); the product is then detected downstream at the second electrode (designated the 'detector' electrode), which is held at a more positive potential, where re-oxidation occurs (E_1; Fig. 1.14d). This mode of operation can offer several advantages over the reductive mode: for certain systems, the magnitude of the applied positive potential at the 'detector' is much lower than the magnitude at the cathode required for reduction; therefore, the background current and consequently the noise is lower. In addition, positive potentials at the 'detector' are usually such that no interference occurs from dissolved oxygen.

The electrode material most commonly employed in commercial amperometric detection systems is glassy carbon; this is resistant to high concentrations of organic solvents, such as acetonitrile and methanol, in the mobile phase. Carbon paste has also been used, usually where aqueous solutions only are used because organic solvents dissolve the pasting agent. Noble metals have also been used in amperometric detectors [7]. In addition, there has been growing interest in the use of chemically modified electrodes in electrochemical detection systems. The present author and coworkers [81] have found phthalocyanine-modified carbon electrodes to significantly reduce the potential required for the oxidation of some biomolecules; this greatly improves the selectivity and sensitivity of detection for these compounds; other chemically modifying agents have been reviewed by Stulik and Pakacova [7].

Improvements in selectivity have also been obtained by using various wave forms applied to the working electrode. Differential pulse measurements were used to effectively filter out an interference with a voltammetric wave preceding that of the analyte [82]; similar studies have been reported with square-wave measurements [83,84]. It should also be added that various pulse-wave forms have been introduced to prevent the adsorption of electrode reaction products and the consequent electrode 'poisoning' [7]. Improvements in resolving overlapping peaks have been obtained by scanning on column, i.e. by rapidly recording a voltammogram on the column eluate. However this technique also reduces the sensitivity due to the presence of capacity current [85]. Another approach has been to use a bank of electrodes, each set at a slightly different potential, and to monitor the sample at all potentials simultaneously; in this way an instantaneous voltammogram is obtained as each analyte elutes. Commercial detectors of his type are available [85].

1.2.2 Coulometric detectors

In coulometric detection an electrochemical reaction occurs at the electrode surface that electrolyses 100% of the analyte [74].

Single-electrode designs have been based on reticulated vitreous carbon, where the eluent containing the analyte flows through the porous electrode rather than over the surface [85–89]. The same principle applies to dual-electrode coulometric detectors; Fig. 1.15 shows a commercially available detector that contains dual porous graphite electrodes in the series configuration.

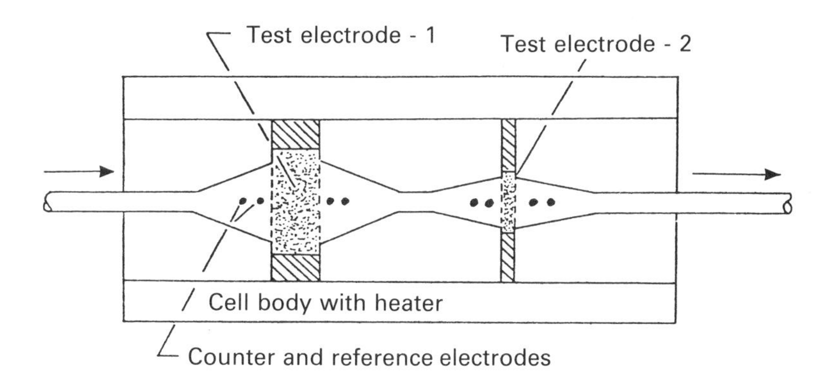

Fig. 1.15 — A dual-electrode coulometric cell, containing porous graphite electrodes in series. (Reprinted by courtesy of Environmental Science Associates.)

The 'generator' electrode and 'detector' electrode areas are about 5 cm^2 and 0.6 cm^2 respectively. With this configuration it is possible to use the redox mode, in the manner described earlier, but in this case total electrolysis occurs at the upstream electrode. A vast improvement in sensitivity was obtained with this, over reductive mode detection, for the detection of vitamin K_1 (Fig. 1.16) [90].

It is also possible to use this electrode arrangement in a different mode, which has been designated the 'screen mode'. The potential of the upstream electrode is set at a value close to the foot of (but not on) the analyte wave, where only the interferences

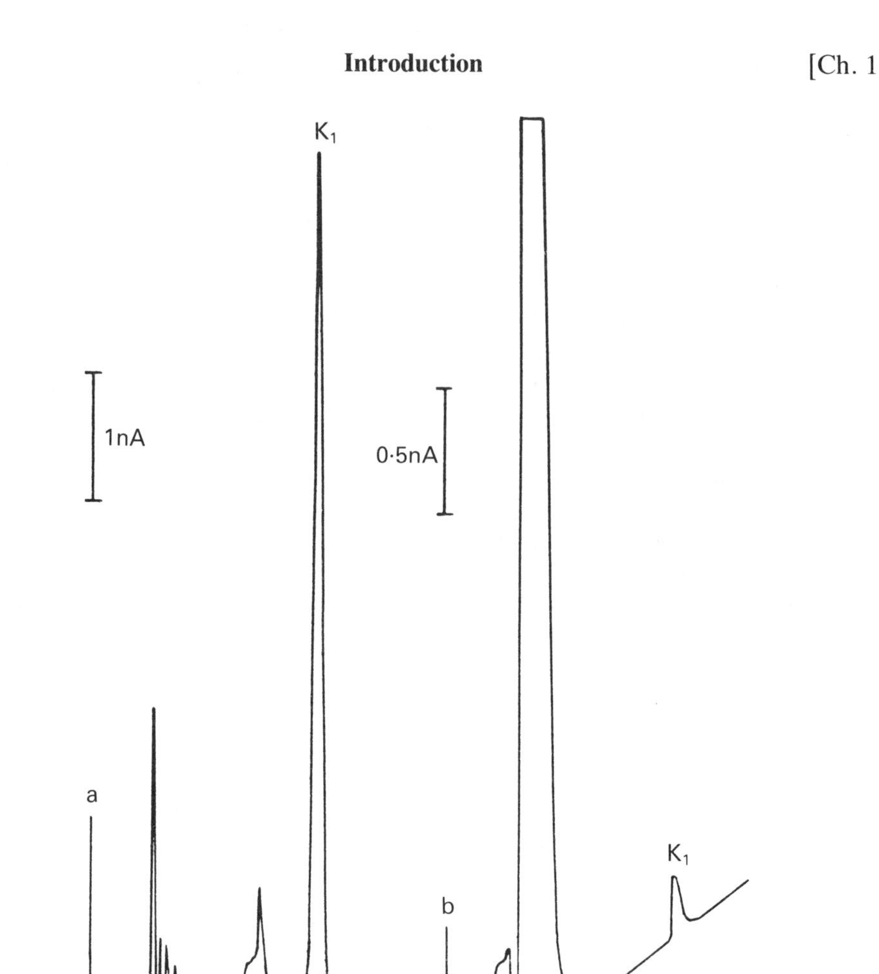

Fig. 1.16 — Chromatograms obtained for 1 ng of vitamin K_1 by LCEC in (a) redox mode and (b) reductive mode. (Reproduced from [90] by permission of the copyright holders, Royal Society of Chemistry.)

are electrolysed. This effectively filters out the potential interference and improves the selectivity of the method; this also offers the possibility of improved sensitivity.

1.3 CONCLUSION

This chapter has described a variety of electroanalytical techniques which may be employed for the study and analysis of biomolecules. The relative sensitivities of these techniques were discussed; of the techniques mentioned, the stripping voltammetric techniques and LCEC allow the lowest detection limits to be achieved. Both of these also possess high selectivity; in LCEC this is achieved by careful choice of the column stationary phase, and mobile phase, as well as the electrode material and configuration. In stripping voltammetry, the selectivity is enhanced by first carrying out the preconcentration step in the sample solution and then transferring the electrode to fresh supporting electrolyte for the measurement step. These features are attractive and may well be the most important criteria in some types of analysis; however, in some cases it may be more desirable to have a short analysis time, in which instance square wave voltammetry, which allows a complete scan in only several seconds, may be more appropriate. In addition to voltammetry and LCEC, amperometric sensors and biosensors can offer simple, and often, rapid, methods for certain types of biomolecules. Clearly, the choice of technique for a particular application depends very much upon the analytical problem. In the following chapters, examples have been included to demonstrate the ways in which the electroanalytical techniques discussed earlier have been used to overcome a wide variety of analytical problems.

REFERENCES

[1] A. M. Bond, *Modern Polarographic Methods in Analytical Chemistry*, Marcel Dekker, New York, 1980.
[2] P. T. Kissinger and W. R. Heineman, (Eds) *Laboratory Techniques in Electroanalytical Chemistry*, Marcel Dekker, New York, 1984.
[3] A. J. Bard and L. R. Faulkner, *Electrochemical Methods*, Wiley, New York, 1980.
[4] B. Fleet and N. Fouzder, in *Polarography of Molecules of Biological Significance*, W. F. Smyth (Ed.), p. 37, Academic Press, London, 1979.
[5] J. G. Osteryoung and R. A. Osteryoung, *Anal. Chem.*, 1985, **57**, 101A.
[6] J. G. Osteryoung, in *Methods in Enzymology*, vol. 158, J. F. Riordan and B. L. Vallee, (Eds), p. 243, Academic Press, San Diego, 1988.
[7] K. Stulik and V. Pacakova, *Electroanalytical Measurements in Flowing Liquids*, Ellis Horwood, Chichester, 1987.
[8] J. Wang, *Electroanalytical Techniques in Clinical Chemistry and Laboratory Medicine*, VCH Publishers, 1988.
[9] J. Heyrovsky, *Chem. Listy*, 1922, **16**, 256.
[10] J. Heyrovsky, *Trans. Faraday Soc.*, 1924, **19**, 785.
[11] P. T. Kissinger, C. R. Preddy, R. E. Shoup and W. R. Heineman, in *Laboratory Techniques in Electroanalytical Chemistry*, P. T. Kissinger and W. R. Heineman (Eds), p. 9, Marcel Dekker, New York, 1984.
[12] G. C. Barker and A. W. Gardner, *Z. Anal. Chem.*, 1960, **173**, 79.
[13] E. P. Parry and R. A. Osteryoung, *Anal. Chem.*, 1965, **37**, 1634.
[14] P. T. Kissinger, in *Laboratory Techniques in Electroanalytical Chemistry*, P. T. Kissinger and W. R. Heineman (Eds) p. 143, Marcel Dekker, New York, 1984.
[15] J. G. Osteryoung and K. Hasebe, *Rev. Polarog. Jap.*, 1976, **22**, 1.
[16] R. N. Adams, *Electrochemistry at Solid Electrodes*, Marcel Dekker, New York, 1969.
[17] W. R. Heineman and P. T. Kissinger, in *Laboratory Techniques in Electroanalytical Chemistry*, P. T. Kissinger and W. R. Heineman (Eds), p.51, Marcel Dekker, New York, 1984.
[18] S. A. Wring, J. P. Hart and B. J. Birch, *Anal. Chim. Acta*, 1990, **231**, 205.
[19] A. P. F. Turner, I. Karube and G. S. Wilson, (Eds) *Biosensors: Fundamentals and Applications*, Oxford University Press, Oxford.
[20] S. Bruckenstein and B. Miller, *Acc. Chem. Res.*, 1977 **10**, 54.
[21] W. J. Alberry and M. L. Hitchman, *Ring-Disc Electrodes*, Oxford University Press, London, 1971.
[22] V. G. Levich, *Acta Physiochim. URSS* 1944 **19**, 133.
[23] V. G. Levich, *Zh. Fiz. Khim.*, 1948, **22**, 711.
[24] F. Vydra, K. Stulik and E. Julakova, *Electrochemical Stripping Analysis*, Ellis Horwood, Chichester, 1976.

[25] W. F. Smyth and I. E. Davidson, in *Electroanalysis in Hygiene, Environmental, Clinical and Pharmaceutical Chemistry*, W. F. Smyth (Ed.), p. 271, Elsevier, Amsterdam, 1980.
[26] J. Wang, *Internatl. Lab.*, 1985, **15**, 68.
[27] W. F. Smyth, in *Electrochemistry Sensors and Analysis*, M. R. Smyth and J. G. Vos (Eds), p. 29, Elsevier, Amsterdam, 1986.
[28] J. Wang, *Anal. Chim. Acta*, 1983, **154**, 87.
[29] J. P. Hart, S. A. Wring and I. C. Morgan, *Analyst*, 1989, **114**, 933.
[30] J. Wang, P. A. M. Farias and S. Mahmoud, *Anal. Chim. Acta*, 1985, **171**, 195.
[31] R. Kalvoda, *Anal. Chim. Acta.*, 1984, **162**, 197.
[32] T. B. Jarbawi and W. R. Heineman, *Anal. Chim. Acta*, 1982, **135**, 359.
[33] E. N. Chaney and R. P. Baldwin, *Anal. Chim Acta*, 1985, **176**, 105.
[34] J. Wang and B. A. Freiha, *J. Electroanal. Chem.*, 1983, **151**, 273.
[35] A. J. Fry and W. E. Britton, in *Laboratory Techniques in Electroanalytical Chemistry*, P. T. Kissinger and W. R. Heineman (Eds), p. 367, Marcel Dekker, New York, 1984.
[36] J. P. Hart and A. Catterall, in *Electroanalysis in Hygiene, Environmental, Clinical and Pharmaceutical Chemistry*, W. F. Smyth (Ed.), p. 145, Elsevier, Amsterdam, 1980.
[37] J. P. Hart, in *Investigative Microtechniques in Medicine and Biology*, vol 1, L. Bitensky and J. Chayen (Eds), p. 199, Marcel Dekker Inc., 1984.
[38] M. A. Brooks and M. R. Hackman, *Anal. Chem.*, 1975, **47**, 2059.
[39] M. A. Brooks, J. A. F. de Silva and L. M. D'Arconte, *Anal. Chem.*, 1973, **45**, 263.
[40] M. R. Hackman, M. A. Brooks and J. A. F. de Silva, *Anal. Chem.*, 1974, **46**, 1075.
[41] J. P. Hart and P. H. Jordan, *Analyst*, 1989, **114,** 1633.
[42] R. W. Murray, in Electroanalytical Chemistry, vol. 13, A. J. Bard (Ed.), p. 191, Marcel Dekker, New York, 1983.
[43] H. G. Barth, W. E. Arber, C. H. Lochmuller, R. E. Majors and F. E. Regnier, *Anal. Chem.* 1986, **58**, 211R.
[44] K. E. Creasy and B. R. Shaw, *Anal. Chem.*, 1989, **61**, 1460.
[45] J. Wang and T. Golden, *Anal. Chim. Acta*, 1989, **217**, 343.
[46] D. Belanger, *J. Electroanal. Chem.*, 1988, **251**, 55.
[47] F. Li and S. Dong, *Electrochim Acta*, 1987, **32**, 1511.
[48] M. Peterssonn, *Anal. Chim. Acta*, 1986, **187**, 333.
[49] P. W. Geno, K. Ravichandran and R. P. Baldwin, *J. Electroanal. Chem.*, 1985, **183**, 155.
[50] S. V. Prabhu and R. P. Baldwin, *Anal. Chem.*, 1989, **61**, 852.
[51] L. M. Santos and R. P. Baldwin, *Anal. Chim. Acta.*, 1988, **206**, 85.
[52] L. Flat and Y. C. Hung, *J. Electroanal. Chem.*, 1983, **157**, 393.
[53] J. Park and B. R. Shaw, *Anal. Chem.*, 1989, **61**, 848.
[54] H. Gomathi and G. P. Rao, *J. Electroanal. Chem.*, 1985, **190**, 85.
[55] M. B. Gelbert and D. J. Curran, *Anal. Chem.*, 1986, **58**, 1028.
[56] L. A. Coury, E. M. Birch and W. R. Heineman, *Proc. Electrochem. Soc.*, 1987, **15**, 104.
[57] L. A. Coury, E. W. Huber, E. M. Birch and W. R. Heineman, *J. Electrochem. Soc.*, 1989, **136**, 1044.
[58] S. A. Wring, J. P. Hart and B. J. Birch, *Analyst*, 1989, **114,** 1563.
[59] G. Jonsson and L. Gorton, *Anal. Lett.*, 1987, **20**, 839.
[60] G. Davis, *Biosensors*, 1985, **1**, 161.
[61] J. E. Frew, S. W. Bayliff, P. N. B. Gibbs and M. J. Green, *Anal. Chim Acta*, 1989, **224**, 39.
[62] H. P. Bennetto, D. R. Dekeyzer, G. M. Delaney, A. Koshy, J. R. Mason, L. A. Razack, J., Stirling and C. F. Thurston, *Analyst*, 1987, **8**, 22.
[63] E. A. Hall, *Enzyme Microb. Technol.*, 1986, **8**, 651.
[64] F. Scheller, F. Schubert, D. Pfeiffer, R., Hintsche, I. Dransfeld, R. Renneberg, V. Wollenberger, K. Riedel, M. Pavlova, M. Kuhn, H. G. Muller, P. M. Tan, W. Hoffmann and W. Moritz, *Analyst*, 1989, **114**, 653.
[65] H. Gunasingham and C. B. Tan, *Analyst*, 1989, **114**, 695.
[66] W. Matuszewski and M. Trojanowicz, *Analyst*, 1988, **113**, 735.
[67] J. B. Flato, *Anal. Chem.*, 1972, **44**, 75A.
[68] M. Porthault, *Electrochemical Detection Techniques in the Applied Biosciences, Volume* 1, *Analysis and Clinical Applications*, Ellis Horwood, Chichester, 1988.
[69] S. A. McClintock and W. C. Purdy, *Internatl. Lab.*, 1984, **14**, 70.
[70] K. Bratin, C. L. Blank, I. S. Krull, C. E. Lunte and R. E. Shoupe Internatl. Lab., 1984, **14**, 24.
[71] P. T. Kissinger, in *Laboratory Techniques in Electroanalytical Chemistry*, P. T. Kissinger and W. R. Heineman (Eds), p. 611, Marcel Dekker, New York, 1984.
[72] D. M. Radzik and S. M. Lunte, *CRC Crit. Rev. Anal. Chem.*, 1989, **20**, 317.
[73] J. P. Hart, in *Electrochemistry Sensors and Analysis*, M. R. Smyth and J. G. Vos (Eds), p. 355, Elsevier, 1986.

[74] B. Hanekamp, P. Bros and R. W. Frei, *Trends Anal. Chem.*, 1982, **1**, 135.
[75] K. Bratin, P. T. Kissinger and C. S. Bruntlett, *J. Liq. Chromatogr.*, 1981, **4**, 1777.
[76] J. M. Elbicki, D. M. Morgan and S. G. Webber, *Anal. Chem.*, 1984, **56**, 978.
[77] J. P. Hart, M. J. Shearer, P. T. McCarthy and S. Rahim, *Analyst*, 1984, **109**, 477.
[78] B. Fleet and C. J. Little, *J. Chromatogr. Sci.*, 1974, **12**, 747.
[79] C. E. Lunte, P. T. Kissinger and R. E. Shoup, *Anal. Chem.*, 1985, **57**, 1541.
[80] R. E. Shoup, *Current Sep.*, 1982, **4**, 36.
[81] S. A. Wring, J. P. Hart and B. J. Birch, *Analyst*, 1989, **114**, 1571.
[82] W. A. MacCrehan, *Anal. Chem.*, 1981, **53**, 74.
[83] R. Samuelsson, J. O'Dea and J. G. Osteryoung, *Anal. Chem.*, 1980, **52**, 2215.
[84] J. Wang, E. Ouziel, C. Yarnitsky and M. Ariel, *Anal. Chim. Acta*, 1978, **102**, 99.
[85] G. M. Greenway, *Chromatogr. Anal.*, 1989, **7**, 9.
[86] W. J. Blaedel and J. Wang, *Anal. Chem.*, 1979, **51**, 799.
[87] A. N. Strohl and D. J. Curran, *Anal. Chem.*, 1979, **51**, 353.
[88] A. N. Strohl and D. J. Curran, *Anal. Chem.*, 1979, **51**, 1045.
[89] D. J. Curran and T. P. Tougas, *Anal. Chem.*, 1984, **56**, 672.
[90] J. P. Hart, M. J. Shearer and P. T. McCarthy, *Analyst*, 1985, **100**, 1151.

2

Pyrimidine and purine derivatives

2.1 INTRODUCTION

Pyrimidine and purine derivatives play a vital role in many biological processes. They exist in nature mainly as components of larger molecules, such as nucleoside and nucleotide constituents; these may be in the free state or present in coenzymes and polynucleotide chains [1]. In the latter case, this may be further incorporated as a component of a nucleic acid, i.e. deoxyribonucleic acid (DNA) or ribonucleic acid (RNA). The genetic information required for the function and multiplication of the biological organism is stored, duplicated and transmitted by means of these nucleic acids [2]. This information is contained in the sequence of bases, whereas sugar and phosphate groups perform a structural role [2].

DNA usually contains four kinds of bases: cytosine (I), and thymine (II) (which are pyrimidines), adenine (III) and guanine (IV) (which are purines). In RNA thymine is mainly replaced by uracil (V) [3]; Fig. 2.1 shows the structures of these five important bases together with the parent compounds pyrimidine (VI) and purine (VII).

The nucleosides are composed of bases to which a sugar is attached; this occurs at the N(9) position in purine bases and at the N(1) position in pyrimidine bases. These nucleosides are, with very few exceptions, either ribonucleosides (containing the sugar ribose) or deoxyribonucleosides (containing the sugar deoxyribose). Since the former possesses three hydroxyl groups on the sugar (positions 2′, 3′ and 5′) and deoxyribonucleosides possess two such hydroxyl groups (positions 3′ and 5′), three and two isomeric monophosphorylated nucleosides, i.e. nucleotides, are respectively possible [4]. The nomenclature used to describe nucleosides and nucleotides is illustrated in Fig. 2.2 using adenine as an example.

The ribonucleoside 5′-monophsophates are generally called adenylic, guanylic, cytidylic and uradylic acids; correspondingly, deoxyribonucleoside 5′-monophosphates are called deoxyadenylic, deoxyguanylic, deoxycytidylic and thymidylic acids [4].

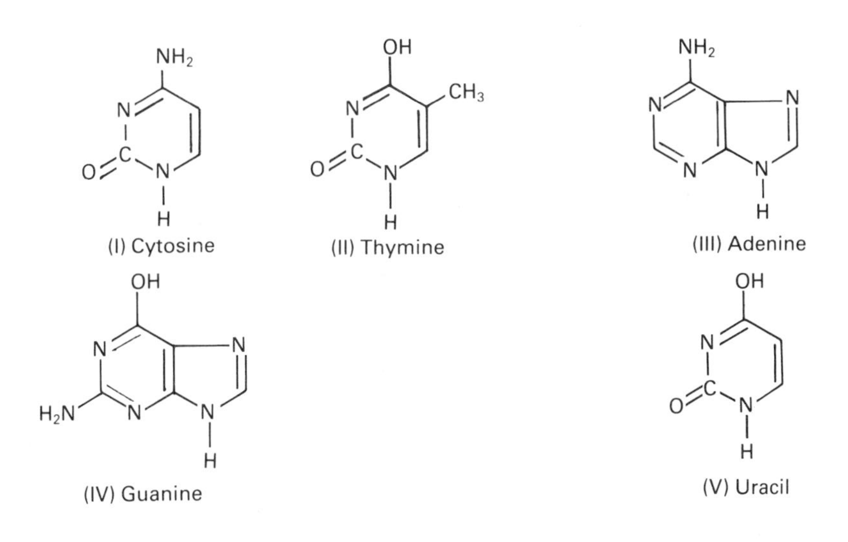

Fig. 2.1 — Structural formulae of the common purine and pyrimidine bases.

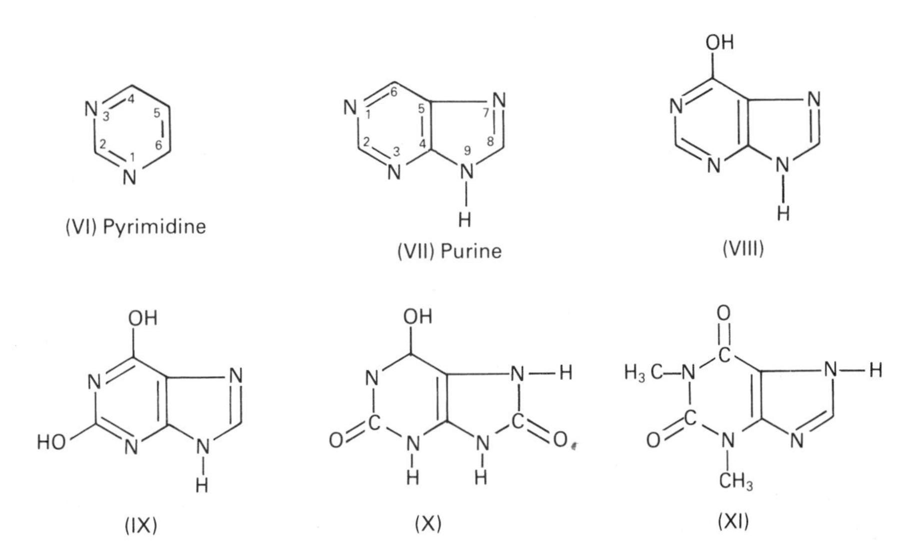

More than 20 minor purine and pyrimidine nucleosides have been found in the hydrolysates of RNA isolated from various organisms [5]. There are several different classes of minor constituents of which methylated bases form the largest class. Several minor components have also been found in DNA in addition to the four common bases. The DNA of animals and plants contains 5-methyl cytosine and that of certain bacteria contains 6-methyladenosine; some other methylated nucleosides, which are minor constituents of nucleic acids, have also been reported [6,7].

The degradation of nucleotides containing adenine and guanine, in some

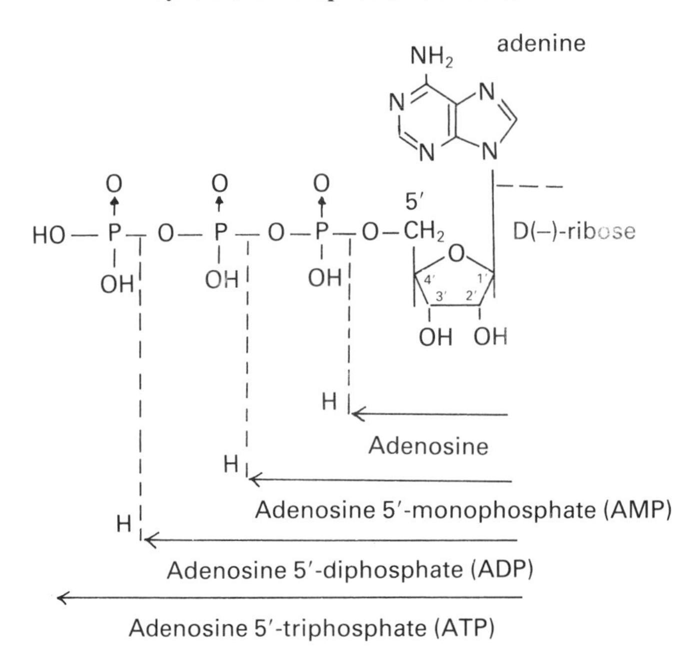

Fig. 2.2 — Formulae for adenine and derived nucleosides and nucleotide species. (Adapted from [4].)

animals, leads to the formation of the oxypurines known as hypoxanthine (VIII), xanthine (IX) and uric acid (X); these are all found in human urine but uric acid is present at much the highest concentrations [8].

Besides the naturally occurring pyrimidine and purine derivatives mentioned above there are also important synthetic substances that contain the purine or pyrimidine ring. In some cases these substances have been shown to possess pharmacological actions, e.g. theophylline (1,3-dimethylxanthine, (XI)) is a widely used bronchodilator used in the management of various asthmatic conditions [9].

From the above discussion it is apparent that the analysis of naturally occurring and synthetic pyrimidine and purine derivatives is of major importance in such areas as clinical chemistry, nucleic acid research and pharmacology. However, these determinations may present difficult analytical problems when the substances are present at very low levels in complex mixtures and/or in biological fluids. But, modern electroanalytical techniques, [10–13] have been used to overcome some of the problems associated with these difficult determinations; this chapter will describe such examples.

This chapter consists of two main sections. The first section describes the electrochemical behaviour of some important pyrimidine and purine derivatives; the second section discusses the electroanalysis of these substances. The latter is divided into subsections according to the electrode material and the technique used in the analysis. These techniques include: polarography; voltammetry at stationary mercury electrodes, which includes linear-sweep and cyclic voltammetry, as well as stripping voltammetry; voltammetry at carbon electrodes, which includes differential-pulse linear sweep and cyclic voltammetry; LCEC with various types of flow-through cell; amperometry with biosensors.

2.2 ELECTROCHEMICAL BEHAVIOUR

2.2.1 Pyrimidine and its derivatives

2.2.1.1 Reduction processes

The electrochemical reduction of pyrimidine and the common bases found in nucleic acids, i.e. cytosine, thymine and uracil, has been investigated in detail by Elving and coworkers [4,14,15,16,17]. These reports indicated that thymine and uracil were not electrochemically reducible under normal polarographic and voltammetric conditions; however, both pyrimidine and cytosine were found to undergo reduction.

The reduction of pyrimidine was studied using the a.c. and d.c. polarographic techniques [4,14,15]; the behaviour of pyrimidine in acetate buffer pH 3.9 is shown in Fig. 2.3 and the overall mechanism of reduction was reported to be that shown in Fig. 2.4. The authors indicated that the number of polarographic peaks and waves which were observed as a result of the reduction of pyrimidine depended on the pH of the supporting electrolyte. It was reported that wave I was only visible in acid solution and was the result of $1e^-$ reduction of the 3,4 bond with simultaneous acquisition of a proton to form a free radical; this may dimerize to 4,4′-bipyrimidine or may be reduced at more negative potentials in a further $1e^-$ process (wave II) to produce 3,4-dihydropyrimidine. Since wave I was strongly pH-dependent and wave II showed only slight pH dependence the two waves were found to merge with increasing pH; at pH about 5 they were replaced by a third wave (III) which was a $2e^-$, pH-dependent process (not shown). In addition, a fourth wave (IV) appeared between pH 7 and 8, which was due to addition of two electrons across the C=N moiety at the 1,3 position. At even higher pH values, i.e. 9 to 13, waves III and IV were replaced by a fifth wave (V); this was the result of the simultaneous addition of four electrons to the two azomethine groups to form a tetrahydropyrimidine.

Similar electrochemical behaviour and conclusions concerning the reduction of pyrimidine were drawn from the cyclic voltammetric behaviour using an HMDE [15]; in this study the authors reported that there was no evidence for reversibility of any of the five peaks. In the same study, it was shown that at a pyrolytic graphite electrode (PGE) only a single cathodic peak appeared over the whole pH range; it was suggested that this corresponds most closely to the succession of peaks I, II and possibly IV observed at the HMDE. Apparently, the lower cathodic potential range of PGE resulted in the absence of peaks III and V; in addition, adsorption was found not to occur in acidic solutions.

The reduction of cytosine (I) at the DME has been discussed in several reports [4,15–17]. It was demonstrated that cytosine exhibited a $3e^-$ d.c. polarographic wave, which occurred at potentials more negative than waves I, II and III of pyrimidine but less negative than wave IV. This was due to the three successive reduction processes shown in Fig. 2.5; the initial reaction was similar to that for pyrimidine in that the two electrons were added across the 3,4-azomethine bond.

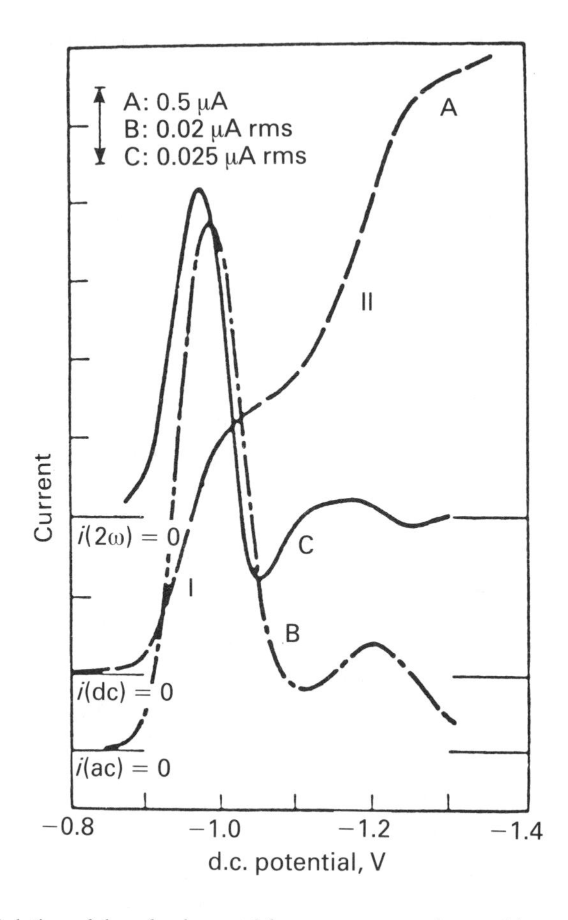

Fig. 2.3 — Relation of d.c., fundamental-frequency a.c., and second-harmonic a.c. polarograms for 0.5-mM pyrimidine in pH 3.9 acetate buffer. (A) d.c. polarogram; (B) Phase-selective a.c. polarogram(in-phase current component), frequency 50 Hz; (C) Phase-selective second-harmonic polarogram, frequency 50 Hz (10 mV rms), response at 100 Hz. Roman numerals refer to the two pyrimidine waves or peaks seen at this pH. (Reprinted with permission from P. J. Elving, S. J. Pace and J. E. O'Reilly, *J. Am. Chem. Soc.*, 1973, **95**, 647. Copyright 1973, American Chemical Society.)

This was followed by chemical reaction with loss of ammonia before further electron transfer took place.

Dryhurst and Elving [15] reported on the cyclic voltammetric behaviour of cytosine at an HMDE and PGE and suggested that there was no evidence of reversibility at either electrode. In the same report, the authors indicated that electrocapillary data (see 4.2.4.1) showed that cytosine was adsorbed onto the mercury electrode (DME); apparently desorption occurred from the electrode prior to the faradaic reaction.

Other cytosine compounds were investigated for reduction at mercury and the following were found to give cathodic waves: cytidine, deoxycytidine, cytidylic acid, deoxycytidylic acid, 5-methyl cytosine, 5-methyldeoxycytidine, and 5-hydroxymeth-

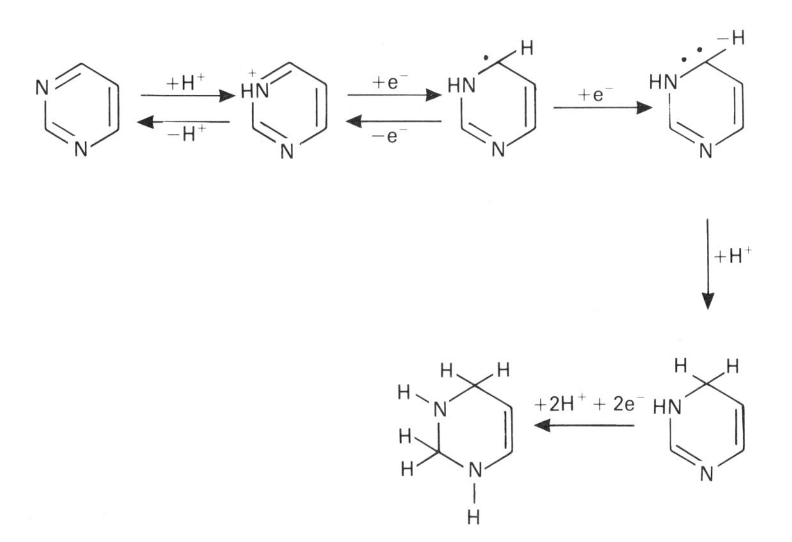

Fig. 2.4 — Mechanism of reduction of pyrimidine. (Adapted from [14].)

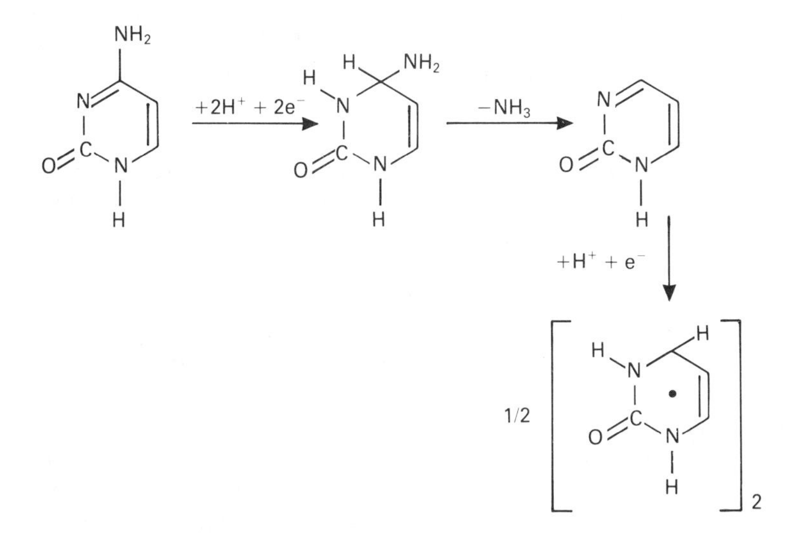

Fig. 2.5 — Mechanism of reduction of cytosine. (Adapted from [4].)

ylcytosine. The waves were generally obtained under acidic conditions. Uracil, thymine and their nucleosides and nucleotides were found not to produce polarographic reduction waves [16,17].

2.2.1.2 Oxidation processes

Investigations using d.c. polarography have been carried out on thymine and uracil and both were found to exhibit anodic waves [18,19]. Apparently, these signals were the result of oxidation of the mercury electrode and proceeded with the formation of sparingly soluble mercury compounds. This phenomenon has been used in conjunction with an HMDE in the development of very sensitive analytical methods involving cathodic stripping voltammetry (CSV); this will be discussed later (section 2.3.2.2).

It has been suggested that the pyrimidine bases do not undergo oxidation at carbon electrodes [20].

2.2.2 Purine and its derivatives

2.2.2.1 Reduction processes

The electrochemical reduction of purine and several important nucleic bases has been studied at mercury and graphite electrodes by Elving and coworkers [4,14,15]. Purine itself was found to be electrochemically reducible, as was adenine; however, guanine did not give any cathodic waves under normal polarographic conditions.

The reduction of purine was studied using a.c. and d.c. polarography; under acidic conditions two peaks or waves (denoted as I and II, for the less negative and more negative signals respectively) were observed [14,15]; these were found to be pH-dependent. The two signals were considered to be the result of two two-electron additions to the 1,6 C=N and the 2,3 C=N, and the overall electrode reaction was found to be irreversible; the reaction described by Elving *et al.* [14] is shown in Fig. 2.6. A.c. polarography of purine, at pH 4.7, indicated that the reaction was accompanied by adsorption at the mercury electrode. The reduction of purine is also complicated by a catalytic hydrogen process which closely follows wave II [14].

Cyclic voltammetry at an HMDE was also used in the same study [14]; at fast scan rates ($24 Vs^{-1}$) an anodic peak was observed. This was considered to be the result of reoxidation of the reduction product formed during wave (I); it was suggested that the product formed during wave (II) was deactivated rapidly by a chemical reaction and, therefore, no reoxidation for this occurred (Fig. 2.6).

Purine was also reported to undergo reduction at a pyrolytic graphite electrode [14,15], but only one cathodic peak was observed. Cyclic voltammetry at the PGE showed that an anodic peak was produced after first scanning to negative potentials to produce peak I (Fig. 2.7); this anodic peak was found to occur over the pH range 1–12. The origin of this peak was suggested to be the same as for the anodic peak found by cyclic voltammetry at the HMDE. The absence of a second cathodic peak was attributed to the smaller potential range available at the PGE. It was reported that the reduction process was also accompanied by adsorption [15].

The reduction of adenine has been reported at mercury electrodes [4,15,21]. It was demonstrated that only one cathodic wave was produced by this purine derivative under acidic conditions, which was reported to be the result of a complex reaction sequence involving an overall six-electron reduction: a $2e^-$ hydrogenation reaction at the 1,6 C=N bond, followed by $2e^-$ reduction of the 2,3 C=N; then, deamination at the 6-position and further $2e^-$ reduction at the regenerated 1,6 C=N bond; finally, hydrolytic cleavage at the 2,3 position occurred to give the same product as the overall $4e^-$ reduction of purine (Fig. 2.8).

Cyclic voltammetry has been carried out on adenine using an HMDE [15]; a single cathodic peak was obtained but no anodic peak was observed even at a scan

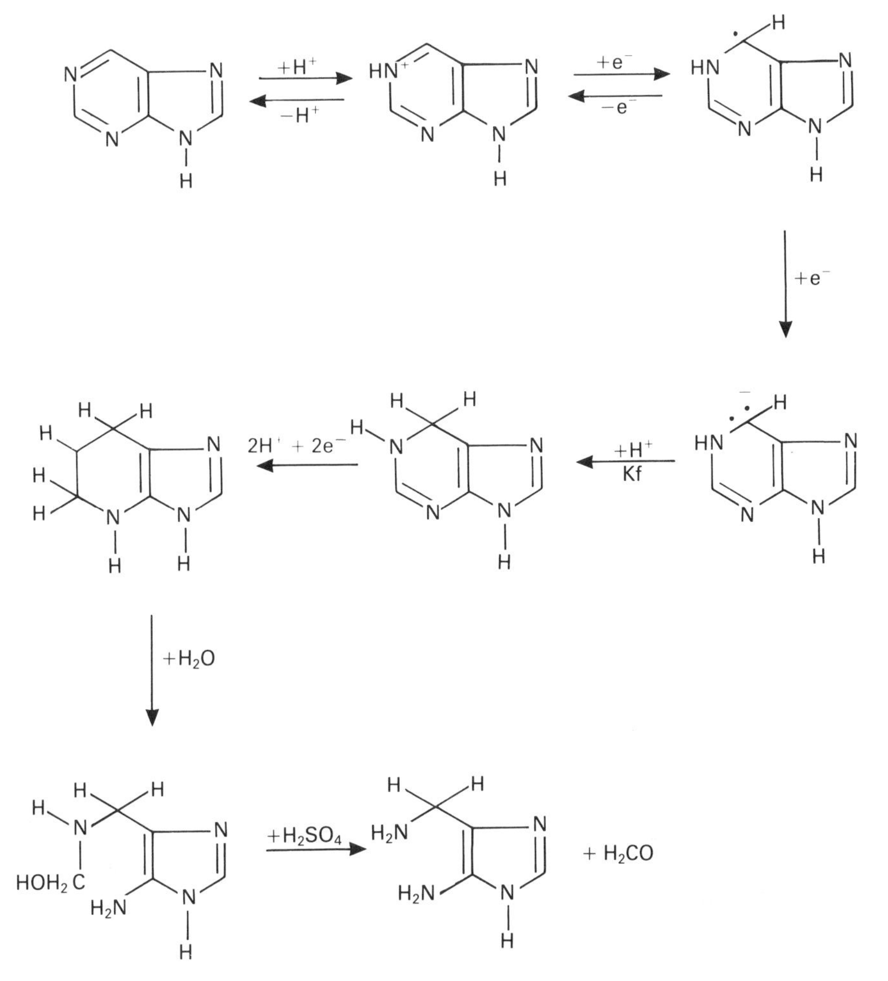

Fig. 2.6 — Mechanism of reduction of purine. (Adapted from [4,14].)

rate of 45 V s^{-1}. The cathodic peak was reported to be pH-dependent between pH 1 and 5.6; the peak current was found to be constant to pH 5.5 and then decreased very rapidly to vanish at pH 6. This peak was considered to correspond to the cathodic wave found at the DME [15].

Adenine did not show any cathodic peaks at a PGE over the pH range 1 to 12; it was suggested that this peak absence was due to the smaller negative potential range available at graphite [15].

The electrochemical reduction of nucleosides and nucleotides containing adenine and guanine bases has been reported by Janik and Elving [4]. It was suggested that compounds containing the adenine moiety gave similar behaviour to adenine itself, i.e. the series of compounds were reduced in the adenine moiety; again it was suggested that only the protonated base was reducible. Guanine nucleosides and nucleotides were reported not to show cathodic signals at mercury electrodes [15].

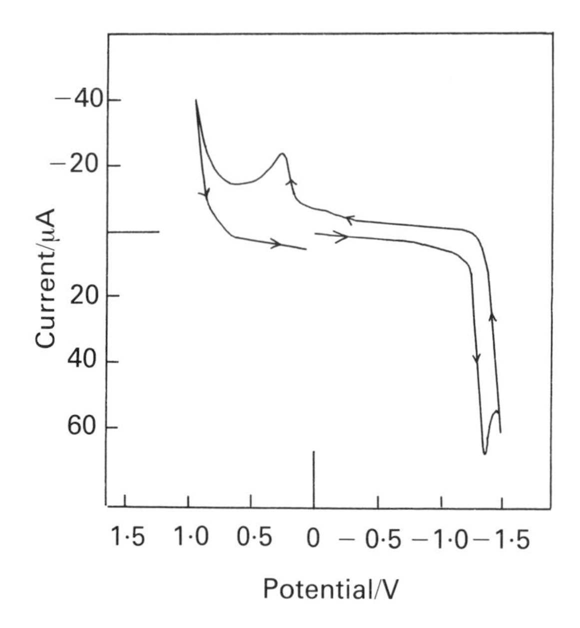

Fig. 2.7 — Cyclic voltammogram at PGE of 1-mM purine in pH 7.0 McIlvaine buffer. Scan rate 60 mV s^{-1}; scan started at 0.0 V and scanned toward negative potentials. (Reprinted with permission from *Talanta*, Vol. 16, Page 855, 1969, G. Dryhurst and P. J. Elving. Copyright 1969, Pergamon Press PLC.)

2.2.2.2 Oxidation processes

The oxidation of purine and some purine derivative has been investigated at mercury electrodes by Palecek *et al.* [22,23]. In this investigation it was shown that purine itself did not produce any anodic signals. However, adenine did yield an anodic wave, as did guanine, xanthine and hypoxanthine and several important methylated adenine derivatives. It was suggested that these were due to oxidation of the mercury electrode with the formation of slightly soluble mercury compounds. As mentioned earlier, this phenomenon has been applied by Palecek *et al.* [22] in the development of very sensitive methods for the analysis of the purine compounds using CSV and will be discussed later (section 2.3.2.2).

There has been considerable interest in the elucidation of oxidation mechanisms of purine derivatives at carbon electrodes; among the reasons given for this is that electrochemical oxidations may mimic biological oxidation mechanisms involving enzymes.

Early studies on the oxidation of guanine at the PGE were carried out by Dryhurst and Pace in 1-M acetic acid [24]; however, it was suggested by Goyal and Dryhurst in a later report [25] that the initial mechanism did not have a great deal of supporting evidence, and that acetic acid may not be a very appropriate medium for their studies. The later study by this group was carried out on guanine over a wider pH range using a combination of electrochemical techniques including LSV, CV, and coulometry at the PGE. In addition, a cell equipped with a reticulated carbon electrode was used for spectroelectrochemical studies [25].

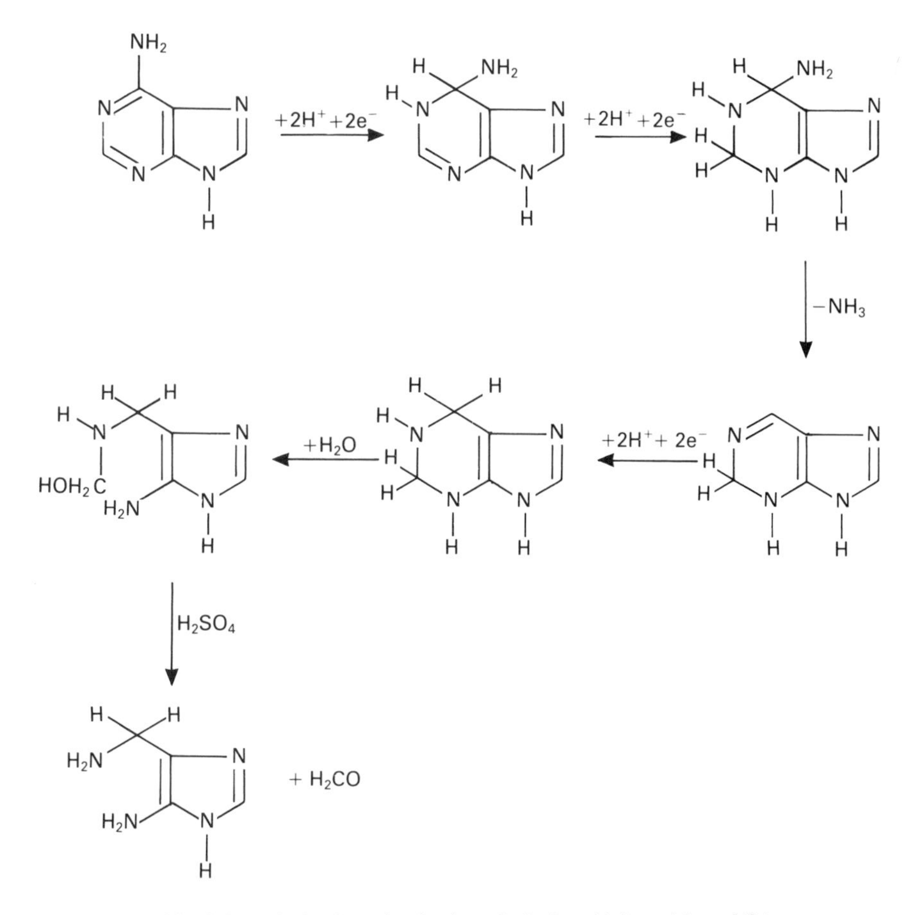

Fig. 2.8 — Mechanism of reduction of adenine. (Adapted from [4].)

These workers [25] reported that LSV and CV of guanine, in phosphate buffers pH 3.3 to 10.3, exhibited a single well-defined oxidation peak (designated as II_a) on the first potential sweep at a clean electrode. Fig. 2.9a, shows the cyclic voltammetric behaviour of guanine at pH 7.0; after scanning beyond peak II_a, three reduction peaks (designated as I_c, II_c and II'_c) were observed on the reverse scan. It was demonstrated that these cathodic peaks only occurred when the anodic scan was first performed. On the second anodic scan a new oxidation peak (denoted I_a) appeared which was said to form an almost reversible couple with peak I_c; in order to observe this anodic peak it was first necessary to scan II_a and then peak I_c. All of the above-mentioned voltammetric peaks were found to be pH-dependent. The oxidation peak I_a, and reduction peaks I_c, II_c and II'_c were also found for 8-oxyguanine as shown in Fig. 2.9b. From these results, and the coulometric and spectroelectrochemical studies, it was concluded that guanine was initially oxidized in an irreversible $2e^-$, $2H^+$ step to give 8-oxyguanine; since this is more easily oxidized, it is immediately electro-oxidized in a further $2e^-$, $2H^+$ process to an unstable quinonoid-diimine

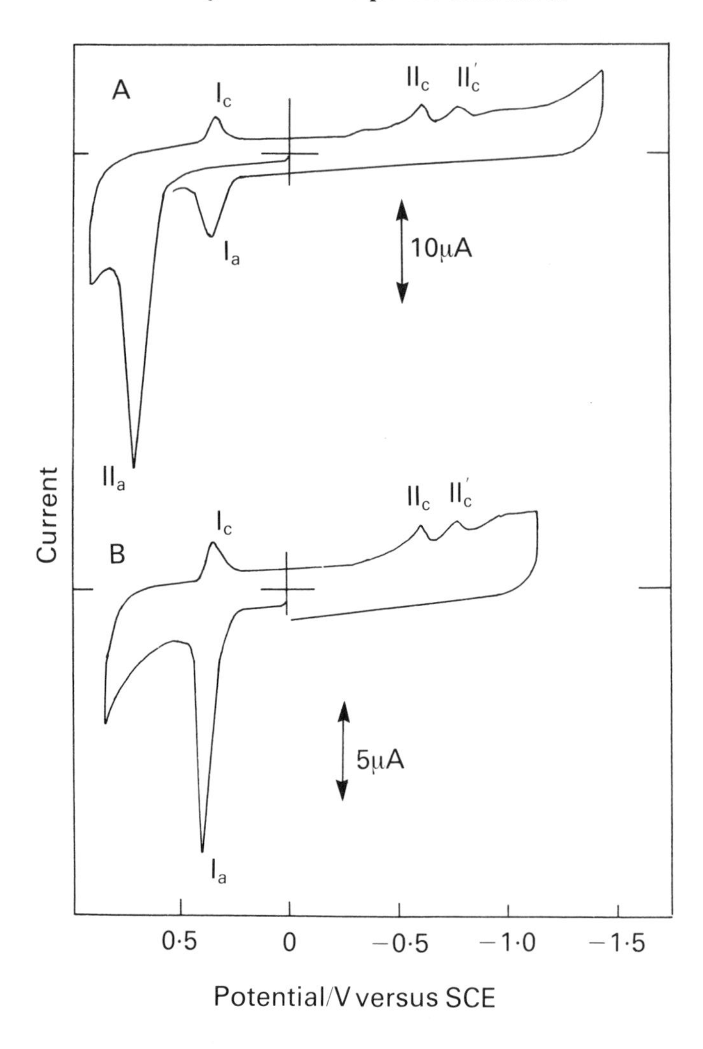

Fig. 2.9 — Cyclic voltammograms at the PGE of: (A) 0.1 mM guanine, and (B) a saturated solution of 8-oxyguanine in phosphate buffer pH 7.0 having an ionic strength of 0.5 M. Scan rate: 200 mV s^{-1}. (Reproduced from [25] by permission of the copyright holders, Elsevier Science Publishers.)

(Fig. 2.10). A series of hydration and other follow-up chemical and electrochemical reactions then occur to give the final products 2,5-diimino-4-imadazolone and 5-guanidino-hydantoin.

The electrochemical oxidation of adenine has also been described; the primary electrochemical oxidation is shown in Fig. 2.11. As may be seen both adenine and guanine were reported to produce a quinoid-diimine; these have been described as unstable species that can undergo further reactions of the type described above. In addition, the metabolic products of these bases, xanthine [26] and hypoxanthine [27], have been studied in detail and were found to undergo similar primary oxidation processes to form unstable quinonoid-diimines.

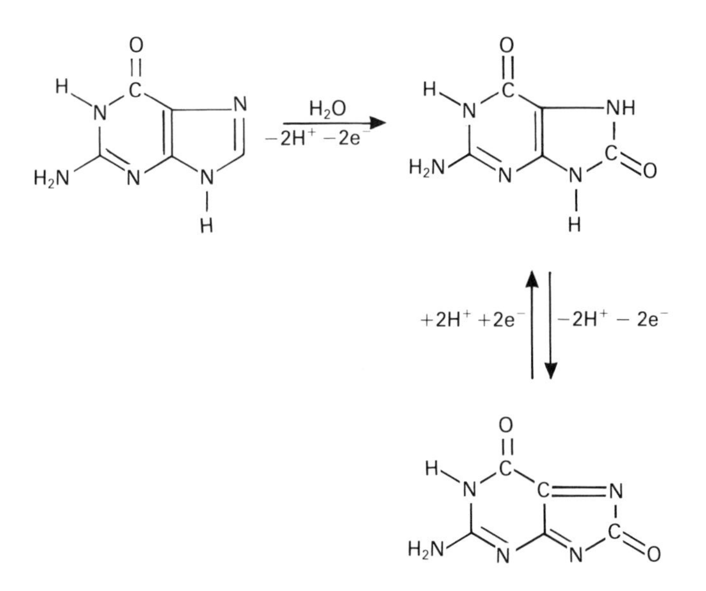

Fig. 2.10 — Primary mechanism of oxidation of guanine. (Adapted from [2].)

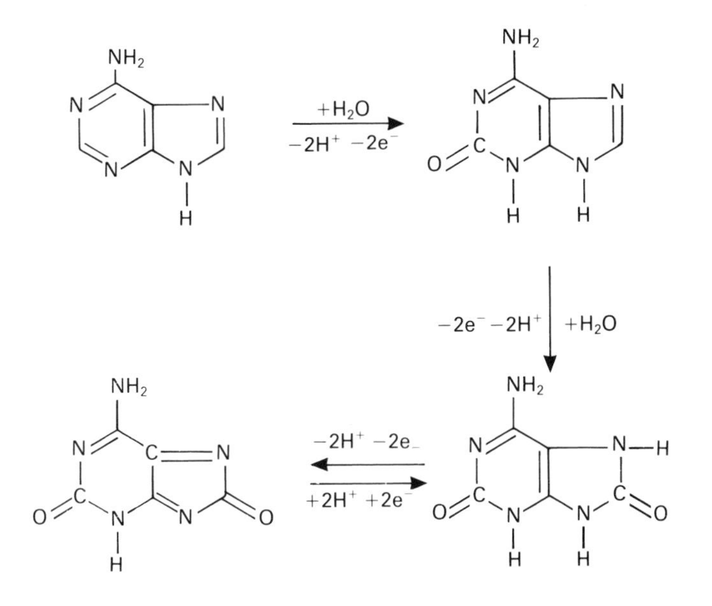

Fig. 2.11 — Primary mechanism of oxidation of adenine. (Adapted from [2].)

The final metabolic product of the purine derivatives in man is uric acid, which is the result of the conversion of xanthine by xanthine oxidase. Dryhurst *et al.* [28,29] has again reported in detail on the electrochemical oxidation of this substance and compared it with a peroxidase enzyme oxidation reaction. Since the molecule exists in different ionic forms depending on pH the possible processes are quite complex (Fig. 2.12, at pH 8.0).

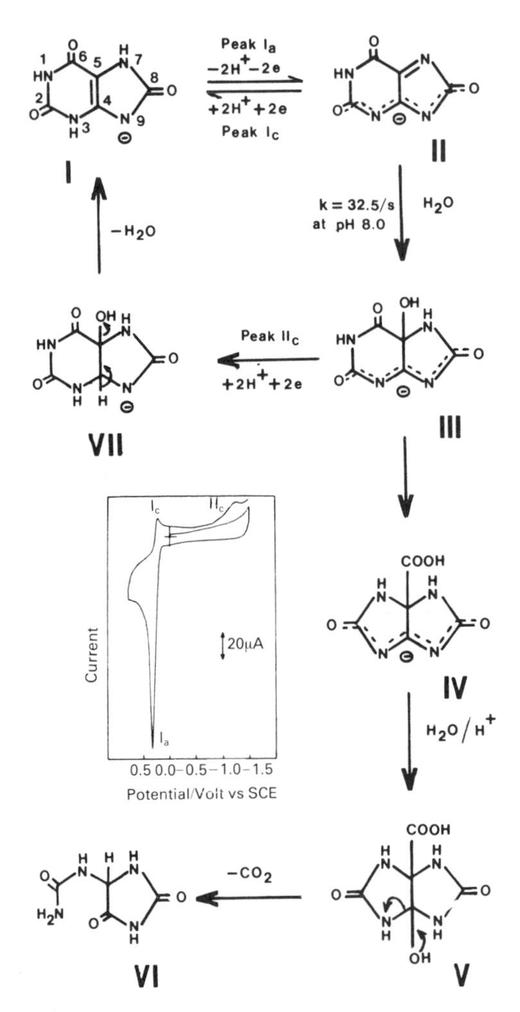

Fig. 2.12 — Reaction mechanism proposed for the electrochemical oxidation of uric acid in phosphate buffers at pH ≥7. (Reprinted with permission from G. Dryhurst, N. T. Nguyen, M. Z. Wrona, R. N. Goyal, A. Brajter-Toth. J. L. Owens and H. A. Marsh, *J. Chem. Ed.*, 1983, **60**, 315. Copyright 1983, American Chemical Society.)

Recently, it has been shown that natural and biosynthetic polynucleotides containing adenine and/or guanine residues are oxidized at PGE and glassy carbon electrodes [2]. In addition it has also been shown that nucleic acids give oxidation peaks at PGE; these apparently coincide with adenine and guanine residues [2].

2.3 ELECTROANALYSIS OF PYRIMIDINE AND PURINE DERIVATIVES

2.3.1 Polarographic methods

Analytical determinations of pyrimidine and purine derivatives, using polarographic methods, were described over 25 years ago by Smith and Elving [30]. In this early study, the workers demonstrated that it was possible to analyse adenine, cytosine and guanine in a mixture containing the three compounds even with the classical d.c.

technique. In the same study, it was also shown that the purines hypoxanthine, xanthine and uric acid could be identified in a mixture containing all three compounds, but the limit of detection for these methods was only about 0.1 mM.

More recently, improved sensitivity for the analysis of pyrimidine and purine compounds has been achieved by the application of DPP. Cummings *et al.* [31] have optimized both the instrumental, and pH conditions for the determination of adenine, cytosine, adenosine, and cytidine alone and in mixtures. The most suitable supporting electrolyte was found to be acetate buffer pH 4.2; in this medium the optimum resolution of adenine from background and adenine from cytosine was achieved. The optimum DPP parameters were found to be: scan rate 2 mV/s; modulation amplitude 25 mV; drop time 1 s. Under these conditions adenine showed a peak at -1.305 V and cytosine exhibited a peak at -1.430 V; therefore, it was possible to measure both compounds in mixtures containing only these two compounds. In the same study it was shown that adenosine and cytidine exhibited peaks at -1.34 V and -1.315 V respectively; although it was not possible to determine these two together in a mixture it was possible to measure cytosine and cytidine in a binary mixture. The determination of these substances in their respective mixtures was studied at concentrations below 10 μM; satisfactory determinations were obtained at the 1-μM level in both cases. In standard solutions of the pure compounds only, the detection limits were: adenine, 0.05 μM; adenosine, 0.04 μM; cytosine, 0.4 μM; cytidine, 0.1 μM.

In a more recent study on cytosine and cytidine mixtures, Temerk *et al* [32], investigated the effect of pH using McIlvaine buffers over a wide pH range (Fig. 2.13a,b). From the results obtained in this study a pH of 5.2 was selected for the analysis of cytosine and cytidine in mixtures. In the same investigation [32] cytidylyl-(3′–5′)-cytidine (Cpc) and cytidylyl-(3′–5′)-adenosine (Cpa) were also studied by DPP (Fig. 2.13c,d); this technique was successfully employed to analyse mixtures containing cytidine and Cpc and also cytidine and Cpa using a pH of 3.2.

Recent studies by Temerk *et al.* [33] have been concerned with the DPP trace determination of several rare nucleic acid components. In this study, 1-methyl-adenine, deoxyadenosine, and adenylyl-(3′–5′)-adenosine were found to exhibit well-defined peaks at pH 3.2, whereas a pH of 4.2 was found more suitable for deoxyadenosine-5-monophosphate and adenylyl-(3′–5′)-guanosine; the detection limits were, 0.9, 0.93, 0.93, 1.80 and 0.91 μM respectively.

DPP was also the technique selected by Kato and Hatano [34] for the determination of low levels of the purine nucleoside inosine, in Britton–Robinson buffer pH 3.0; the calibration graphs were found to be linear up to a concentration of 100 μg/ml.

A detailed investigation has been carried out by Bouzid and Macdonald [35] on 18 uracil derivatives using a variety of electroanlytical techniques. The substances studied included the antineoplastic agent 5-fluorouracil as well as antiviral agents which consisted of pyrimidine compounds substituted at the 5-position. These authors showed that DPP could be used for the determination of these substances by exploiting the anodic wave produced by the formation of sparingly soluble mercury salts, Table 2.1 shows the compounds studied and the calibration data obtained by DPP. It was stated that differences in the range of linearity may reflect the different areas occupied on the electrode surface. Compounds such as uracil and 5-azauracil

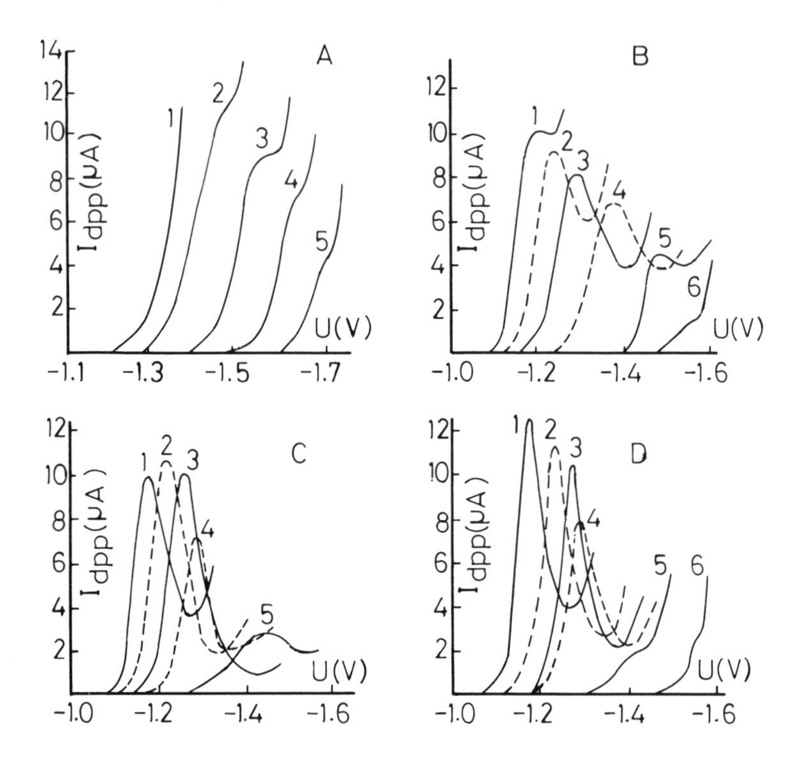

Fig. 2.13 — Effect of pH on the differential pulse polarograms of: (A) 5×10^{-5}-M cytosine; (1) pH 3.2, (2) 4.2, (3) 5.2, (4) 6.2, (5) 7.2); (B) 5×10^{-5}-M cytidine; (1) pH 2.2, (2) 2.2, (3) 3.2, (4) 4.2, (5) 6.2, (6) 7.2,; (C) 5×10^{-5}-M Cpc; (1) pH 3.2, (2) 4.2, (3) 4.8, (4) 5.2, (5) 6.2; (D) 5×10^{-5}-M Cpa; (1) 3.2, (2) 4.2, (3) 4.8, (4) 5.2, (5) 6.2, (6) 7.2. Scan rate 2 mV s^{-1}, drop time 2 s, pulse amplitude 100 mV, *T* 22°C. (Reproduced from [32] by permission of the copyright holders, Elsevier Science Publishers.)

may be adsorbed flat on the electrode surface, whereas substances containing bulky substituents such as 5-iodouracil and 5-nitrouracil may have to reorientate from the planar position prevailing at low concentration to a perpendicular one at high concentrations.

Much interest has been shown in the application of polarographic methods to studies concerned with native (double-helical), and denatured (single-stranded), nucleic acids as well as synthetic polynuclotides. Palecek [36] discussed the use of alternating current polarography in studies related to the adsorption of nucleic acids and polynucleotides at the mercury surface. Tensammetric (non-faradaic) peaks (Fig. 2.14) were reported to appear for DNA at neutral pH and at moderate ionic strength; the potentials required for this phenomenon to occur were between about 0 and −1.0 V. The appearance of peak 1 (at −1.2 V), has been explained as being due to segmental desorption of helical regions, which are first adsorbed via the sugar phosphate backbone in both native and denatured DNA [36,37]; peak 2 is exhibited only by native DNA and it was suggested that this might be the result of adsorption/desorption properties associated with open double-stranded regions [36,38]. The

Table 2.1 — Characteristics of the calibration plots for pyrimidine derivatives

Substance	Conc. range (10^{-5} M)	Equation[a]	Correlation coefficient (r)[b]	Standard error
Uracil	0.5–40	$y=0.738x+0.119$	0.997	0.1346
Thymine	0.5–10	$y=0.736x+0.156$	0.9995	0.1023
5-Fluorouracil	0.5–10	$y=0.750x+0.055$	0.999	0.0517
5-Trifluoromethyluracil	1.0–10	$y=0.691x+0.071$	0.999	0.0581
5-Chlorouracil	0.5–10	$y=0.740x+0.092$	0.999	0.1042
5-Bromouracil	0.5–10	$y=0.437x+0.090$	0.999	0.0948
5-Iodouracil	0.5–10	$y=0.321x+0.132$	0.992	0.1769
5-Acetyluracil	0.5–20	$y=0.692x+0.043$	0.995	0.1980
5-Formyluracil	0.5–10	$y=0.661x+0.151$	0.998	0.2089
5-Vinyluracil	0.5–10	$y=0.514x+0.011$	0.999	0.0431
5-Nitrouracil	0.5–2	$y=0.374x+0.019$	0.997	0.0290
5-Ethynyluracil	0.5–8	$y=0.451x+0.077$	0.997	0.1112
5-Azauracil	0.5–40	$y=0.137x+0.047$	0.999	0.0390
6-Azauracil	1.0–10	$y=0.358x+0.046$	0.997	0.0638
6-Chlorouracil	0.5–8	$y=0.182x+0.065$	0.970	0.1737
6-Methylthymine	1.0–10	$y=0.640x+0.113$	0.999	0.0921
2-Thiouracil[c]	0.5–8	$y=0.308x+0.105$	0.993	0.1317

[a]Units: y, measured current (μA); x, concentration (10^{-5} M).
[b]$n=5$–7.
[c]Determination at pH 12.2.
(Reproduced from [31] by permission of the copyright holders, Elsevier Science Publishers Physical Sciences and Engineering Division.)

third peak (peak 3) was reported only for denatured DNA; it was suggested that this was the result of desorption of firmly adsorbed dbases from single-stranded regions of polynucleotides [36,38]. Under the same conditions, and at ionic strengths less than 0.2 M, it was shown that double-helical DNA gave a peak 0; it was suggested that this was related to electrostatic interactions of native DNA at the mercury surface [36,38].

DPP of nucleic acids shows similar current–voltage curves to those obtained by the a.c. technique; the former has been used in a variety of investigations on DNA. Single-stranded DNA produces a well-defined DPP peak 3 at neutral pH, whereas native DNA does not; when pulse amplitudes of 50 or 100 mV are employed concentrations of denatured DNA can be determined down to 0.1 μg/ml [39]. Since the double-helical nucleic acids do not exhibit peak 3, traces of single-stranded DNA can be determined in samples of native DNA and RNA [40]. Other electrochemical studies have also been carried out on DNA using peak 3. Modifications of double-helical DNA caused by the action of alkali and heating causes 'melting' and formation of single strands of DNA at the end of the helix; consequently, DPP of DNA after subjection to these treatments results in the appearance of peak 3 [38].

Further DPP studies have utilized peak 2 and it was possible to use this peak to monitor changes in the local structure of double helical DNA. The studies carried out included: the investigation of photoproducts caused by low doses of UV radiation [41]; the effect of heating to premelting temperature, which causes a premelting type of structure [38]; and the effect of low doses of x-rays and γ-rays, which cause single-stranded DNA to break [41]. In these cases the magnitude of peak 2 was found to decrease.

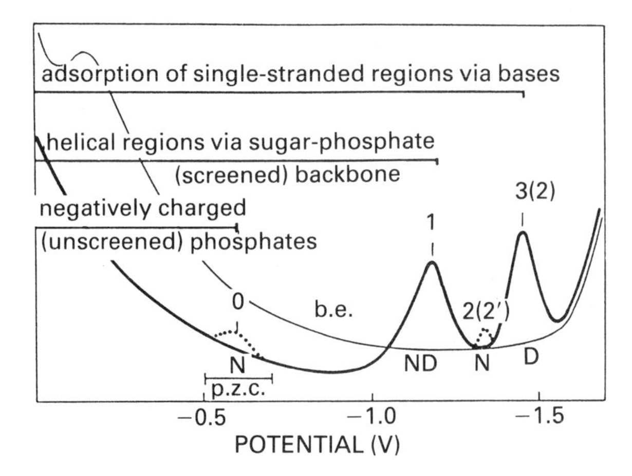

Fig. 2.14 — a.c. polarograms of native and denatured DNA and of the possible role of DNA constituents in its adsorption at DME. N, a peak only produced by double-helical (native) DNA; D, a peak produced only by single-stranded (denatured) DNA; ND, a peak produced by both native and denatured DNA. . . . peaks which appear only under special conditions; b.e., background electrolyte; p.z.c., potential of zero charge. Peaks in brackets are old method of notation. (Reproduced from [36] by permission of the copyright holders, Elsevier Science Publishers.)

2.3.2 Voltammetry at stationary mercury electrodes

2.3.2.1 Linear-sweep voltammetry (LSV) and cyclic voltammetry (CV)

Linear-sweep and cyclic voltammetric techniques, using stationary mercury electrodes, have made significant contributions in nucleic acid research. Investigations have included studies to determine the changes in conformation of nucleic acid structure, under various solution conditions, and at various potentials. In addition, studies have been carried out to assess the changes that occur in DNA structure when it has been exposed to genotoxic agents. The voltammetric studies have generally been based on the measurement of signals produced by oxidation–reduction reactions of the residual bases in the DNA structure, which are accompanied by adsorption phenomena; also non-faradaic signals, produced as a result of adsorption/desorption behaviour, have been used. In this subsection a few examples will be given to demonstrate the way in which LSV and CV techniques have been used in these broad research areas. However, for more detailed surveys the reader is referred to several excellent reviews [2,36,42,43].

LSV and CV at an HMDE were employed by Palacek and coworkers [2,36,44,45,46] in studies involving the surface denaturation of DNA at different pH values.

The cyclic voltammetric behaviour of native and denatured DNA has been described by Palecek [36]; cyclic voltammograms obtained by this worker, in neutral solution, are shown in Fig. 2.15. The faradaic currents (peak III) obtained for denatured DNA were reported to be independent of E_i (the initial potential prior to scanning). However, when solutions containing native DNA were held at E_i around

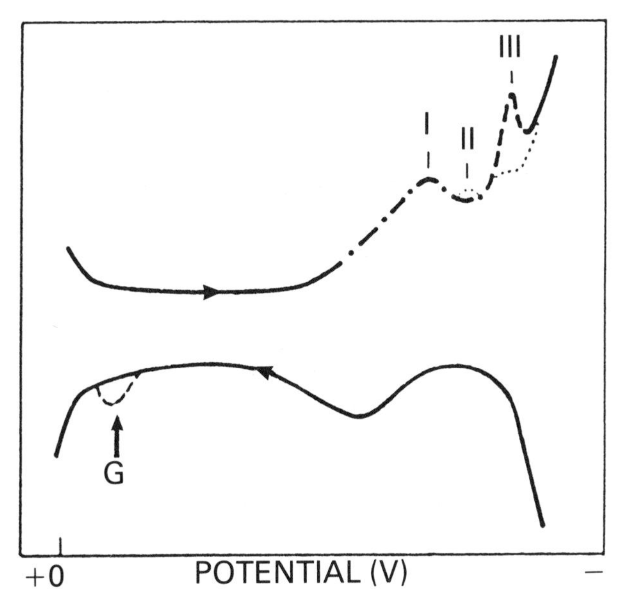

Fig. 2.15 — Cyclic voltammograms obtained for native and thermally denatured DNA at neutral pH starting at E_i+0.0 V. ———, peaks of denatured DNA, . . ., peak of native DNA (peak III is substantially smaller than in denatured DNA, peak II is produced only by native DNA). (Reproduced from [36] by permission of the copyright holders, Elsevier Science Publishers.)

−1.2 V (for 10 s) peak III was found to increase significantly in height. The explanation for this was that changes in DNA conformation similar to denaturation had occurred; this was supported by the observation that peak II, which only appears on voltammograms of native DNA, had disappeared. The mechanism for denaturation was suggested to be adsorption at mercury via bases at a single-stranded segment, while the adjacent segments are strongly electrostatically repulsed from the negatively charged electrode; therefore, DNA unwinding was said to occur. Following this, additional adsorption could then occur via newly exposed bases and consequently further unwinding of DNA could take place.

Under acidic conditions (pH 5.1), it was shown that the faradaic current for double-stranded DNA was a maximum for E_i from about −0.4 V to −1.0 V and the current actually decreased at E_i about −1.2 V [43]; the expanation given for this observation was that destabilization of the nucleic acid structure had occurred owing to the protonation of bases; adsorption onto the mercury electrode was via these bases. In contrast, it was shown by Palacek [46] that in alkaline solutions neither native nor denatured DNA gave faradaic currents, but capacitance currents were produced instead. Native DNA showed three voltammetric peaks and peak (III) was again dependent on E_i; denatured DNA showed two peaks, peak (I) and peak (III), and the latter was again independent of E_i. From the voltammetric results obtained for native DNA, it was suggested that the mechanism of surface denaturation in alkaline medium was similar to that postulated in a neutral medium [42].

Further work on nucleic acid behaviour using LSV at stationary mercury electrodes has been reported by Sequaris [43]. This author also reported that DNA exhibited cathodic faradaic currents between pH 4 and 7, which were said to be the result of the reduction of accessible adenine and cytosine residues; under acidic conditions (pH 5.6) strongly adsorbed products were formed. The proposed interfacial structure reported in this paper was that the bases were orientated perpendicular to the electrode surface with sugar/phosphate backbone orientated towards solution. Sequaris [43] has also investigated the voltammetric behaviour of DNA at other pH values and has given an explanation for the possible mechanisms of adsorption at mercury; he has also indicated the importance of such findings in nucleic acid research.

Additional voltammetric studies by Sequaris and coworkers have addressed the problem of assessing the effect of genotoxic agents such as radiation damage, chemical agents and ultrasound on nucleic acid structure; this may have possible mutagenic and carcinogenic consequences [42,43,47].

Sequaris and Valenta [42] demonstrated that LSV at an HMDE could be applied to the assessment of γ-radiation damage in native DNA. In these studies DNA from calf thymus and from bacteria was employed; it was shown that slightly acidic or alkaline solutions could be used for the *in vitro* study. In one method, the conditions used were as follows: the supporting electrolyte was McIlvaine buffer pH 5.6; the sweep rate was 5 V s^{-1}; E_i was -1.2 V; the adsorption time at -1.2 V was 16 s; the DNA was from calf thymus. The samples of DNA were exposed to γ-radiation (up to 36 krad) and were then analysed using the voltammetric conditions given above; the voltammetric currents were found to increase with increased doses of γ-radiation. The results obtained were comparable to those obtained by the conventional sedimentation techniques.

Recently, Palacek *et al.* [48] proposed the use of CV to investigate the behaviour of native DNA following irradiation. In addition to following changes in the cathodic peaks, these workers showed that an anodic peak could also be used to probe alterations in the structure. This anodic peak was reported to be due to guanine residues that were accessible after radiation damage (incidently, the anodic guanine peak had been shown on an earlier cyclic voltammogram of DNA, Fig. 2.15 [36]). For these studies [48] a supporting electrolyte containing 0.6 M ammonium formate with 0.1 M phosphate at pH 6.9 was employed; other conditions included: scan rate 200 mV s^{-1}; E_i–0.1 V for 10 s; switching potential -1.85 V; concentration 100 μg ml^{-1}. Following exposure to small doses of γ-radiation the solutions containing DNA showed an increase in both anodic and cathodic voltammetric currents; however the latter were always greater (Fig. 2.16). It was concluded that, under the doses of radiation investigated, the DNA double-helical structure was mainly damaged in the adenine–thymine-rich regions.

Sequaris *et al.* [47] have also investigated the effect of chemical agents on the voltammetric behaviour of calf thymus DNA; this was carried out *in vitro*. These authors showed that there was an increase in the voltammetric signal in the presence of a methylating agent; this indicated that some of the base pairs in native DNA had been labilized. Therefore, it would appear that it may be possible to use stationary mercury electrode voltammetry as a means of screening for potentially carcinogenic/mutagenic chemicals by *in vitro* investigations on DNA [38]. No doubt this promising screening procedure will be further exploited in future research studies.

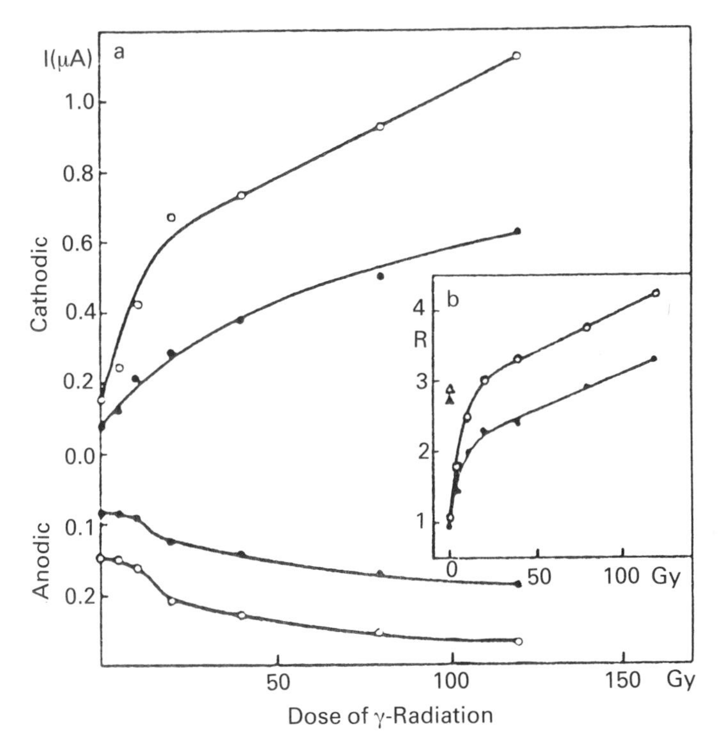

Fig. 2.16 — Relationship between the height of the cathodic (adenine and cytosine) and anodic (guanine) CV peaks of double-stranded DNA and the γ-radiation dose, R, height ratio of the cathodic and anodic peaks ($I_{cathodic}/I_{anodic}$), ○–○ waiting period 10 s (electrode not fully covered); ●–● waiting time 120 s (adsorption equilibrium) at a potential of −0.1 V. R in thermally denatured DNA: △, waiting time 10 s and ▲, 120 s; concentration 100 μg cm^{-3}, HMDE 2.2 mm^2; buffer 0.6-M ammonium formate with 0.1 M sodium phosphate (pH 6.8); scan rate 200 mV s^{-1}. Initial potential −0.1 V, switching potential −1.85 V, waiting time 10 s (at −0.1 V). (Reproduced from [48] by permission of the copyright holders.)

2.3.2.2 Stripping voltammetry

In recent years, Palacek and coworkers have investigated the possibility of measuring very low levels of purine and pyrimidine derivatives by stripping voltammetric procedures at stationary mercury electrodes.

In one study [22], the workers investigated the cathodic stripping voltammetric behaviour of 30 purine and pyrimidine derivatives. As shown in Table 2.2 the majority of the substances studied could be determined by CSV. The authors found that the optimum deposition potentials for the compounds studied were slightly positive; the stripping peaks were close to these values (Table 2.3). In this investigation, the mechanism of preconcentration was reported to be via the formation of a sparingly soluble mercury compound that was adsorbed onto the electrode; this was apparently formed during the oxidation process. The stripping step was carried out either by the application of a fast linear potential ramp or by a differential pulse waveform. The supporting electrolytes used in the study consisted of borate buffer prepared from 0.05-M boric acid adjusted to the desired pH with 0.05-M potassium hydroxide (Tables 2.2 and 2.3). The detection limits obtained for these compounds were in the range 10^{-8} to 10^{-9} M, with greater sensitivity for the purines.

Table 2.2 — CSV of pyrimidine and purine derivatives

Substance	CSV response
Pyrimidine derivatives	
Uracil	+
Cytosine	+
Thymine	+
5-Methylcytosine	I
5-Hydroxymethylcytosine	I
Orotic acid	+
2-Amino-4,6-dioxypryimidine	+
4-Amino-2,6-dioxypryimidine	+
5-Amino-2,4-dioxypryimidine	+
2-Aminopyrimidine	—
2-Hydroxypyrimidine	—
2-Amino-4,6-dioxy-5-methylpryimidine	+
Isocytosine[a]	+
Uracil-5-carboxylic acid	+
Purine derivatives	
Purine	—
Adenine	+
6-Benzylaminopurine	+
Hypoxanthine	+
6-Hydroxymethylpurine	+
2-Aminopurine	+
2-Hydroxypurine	+
2-Hydroxy-6-methylpurine	+
Xanthine	+
Isoguanine	+
8-Hydroxyadenine	+
Uric acid	+
6-Methylpurine	—
Caffeine	—
Theobromine	—
Theophylline	—
7-Hydroxymethyltheophylline	—
Aminophylline	—

+, stripping peak; I, inflexion; —, neither peak nor inflexion appeared under given conditions.
Purine and pyrimidine derivatives at a concentration of 5×10^{-5} M in borate, pH 10.5. Linear potential sweep CSV: deposition potential +0.1 V (for pyrimidines) or +0.07 V (for purines), deposition time 30 s (without stirring), HMDE surface 1.8 mm^2, scan rate 0.2 V s^{-1}.
[a]Measured at a concentration of 5×10^{-5} M.
(Reproduced from [22] by permission of the copyright holders, Elsevier Science Publishers.)

In a separate study Palacek *et al.* [23] examined the CSV behaviour of adenine and its methylated derivatives, 1-methyl-6-aminopurine (1MeAde), 3-methyl-6-aminopurine (3MeAde), *N*-methyl-6-aminopurine (6MeAde) and *N,N*-dimethyl-6-aminopurine ($6Me_2Ade$). All of these, except the last compound gave cathodic stripping peaks due the formation of slightly soluble compounds with mercury. It was suggested that the 6-amino group of adenine was the mercury binding site in these

Table 2.3 — Potentials (U_p) and half-widths ($W_{1/2}$) of CSV peaks of some purine and pyrimidine derivatives

Substance	U_p (V)	$W_{1/2}$ (mV)
Adenine	−0.01	32
Xanthine	−0.07	45
Hypoxanthine	+0.04	12
	−0.06	45
Isoguanine	−0.03	50
	−0.15	40
Uric acid	−0.07	50
Uracil	−0.03	35
Cytosine	0.00	32
Thymine	−0.01	48

Measurements were performed in borate buffer, pH 10.1 at a concentration of substances of 3×10^{-7} M. Linear potential sweep CSV: deposition potential +0.08 V, deposition time 2 min, 20 s quiescent period, scan rate 0.2 V s^{-1}, HMDE 2.9 mm^2.
(Reproduced from [22] by permission of the copyright holders, Elsevier Science Publishers.)

cases. The CSV behaviour of 1MeAde at different deposition times is illustrated in Fig. 2.17; 3MeAde showed similar behaviour. However, 6 Me_2Ade exhibited different CSV characteristics and two stripping peaks (1) and (2) were obtained at concentrations less than 0.057 mM; at 0.107 mM a third peak (3) was formed that masked the other two (Fig. 2.18). The buffer used was borax pH 9.2. The sensitivity (peak height) of peak 1 was found to be about 10 times that for peak 2; the detection limit for this compound was shown to be around 10^{-9} M.

Bouzid and Macdonald have investigated the cathodic stripping behaviour of a number of uracil derivatives [35,49]. They carried out preliminary studies using CSV under non-flowing conditions on 5-flurouracil and 5-vinyluracil; under the optimized conditions both compounds gave linear calibration graphs in the range 0.5–5×10^{-7} M. Following this the authors [49] used a method involving flow-injection differential pulse CSV and extended the number of uracil compounds investigated to include most of those given earlier in Table 2.1. In this investigation a mobile phase containing borax/KNO_3/HNO_3 (or NaOH), with 0.001% (v/v) Triton X-100, was found most suitable; generally the pH was adjusted to a value of 7.6. In this mode, a potential (E_d) which was optimally more positive than the peak potential [31] was applied well before the sample front reached the cell. This potential was kept constant for a period of time depending on the flow rate until the sample passed the electrode; the flow was then stopped and the stripping was initiated after a quiescent period of 15 s by scanning towards negative potentials. The effect of deposition potential, flow rate and sample size on the analytical signals was investigated. The calibration graphs obtained on the above mentioned compound were typically linear in the ranges about 1×10^{-7} to 1×10^{-8} M; for 5-iodouracil the limit of detection was 5×10^{-9} M.

CVS has also been applied to other studies of purine and pyrimidine derivatives [50–57].

As mentioned previously, CSV involves accumulation of the electroactive substance onto the electrode surface by a faradaic reaction. However, adsorptive

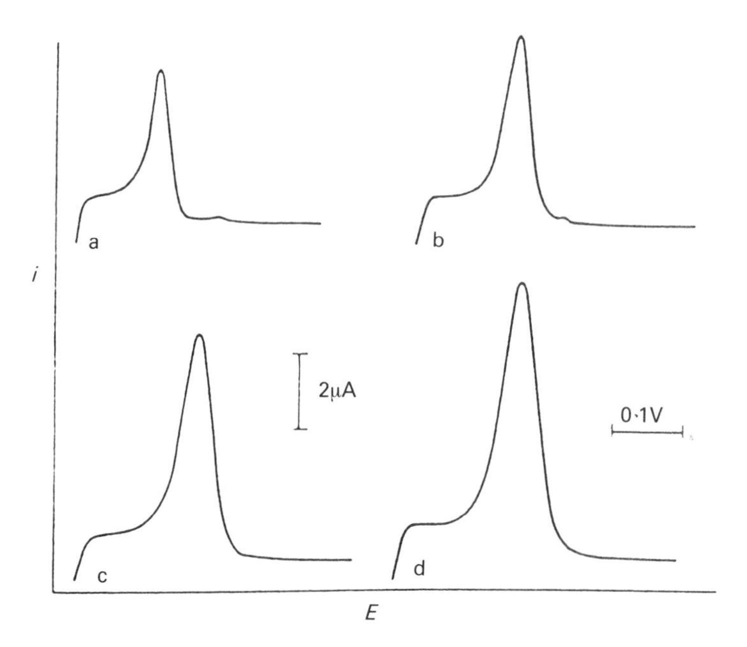

Fig. 2.17 — Dependence of CSV response for 1-mM 1 MeAde on deposition time, t_d. t_d=(a) 10, (b) 30, (c) 60, and (d) 120 s. *Ed* was +155 mV; buffer 0.05 M borax, pH 9.2; scan rate 25 mV s^{-1}; HMDE. (Reprinted with permission from E. Palecek, J. G. Osteryoung and R. A. Osteryoung, *Anal. Chem.*, 1982, 54, 1389. Copyright 1982, American Chemical Society.)

stripping voltammetry (AdSV), in which there is no faradaic reaction during accumulation, has recently been applied to low-level determinations of nucleic acids.

Palecek and Hung [58] investigated the conditions necessary to optimize the adsorptive preconcentration of osmium-labelled DNA and also the differential-pulse cathodic-stripping step. The authors stated that osmium was introduced into the polynucleotide chain as an electroactive marker; it was suggested that this approach may yield valuable information concerning DNA structure. It was shown that the sensitivity for single-stranded DNA was much greater than for native DNA; the former could readily be measured at the 5 ng ml^{-1} level. The supporting electrolyte consisted of ammonium formate/sodium phosphate, pH 6.8, and under these conditions a stripping peak was found at −1.2 V. The differential-pulse waveform was shown to give a larger peak than the LSV technique.

In further investigations by Palecek *et al.* [59] adsorptive preconcentration at mercury was examined for the determination of unmodified nucleic acids, particularly DNA. It was reported that cyclic voltammograms of DNA exhibited an anodic peak at about −0.3 V after first sweeping to cathodic potentials (−1.85 V); this was considered to be produced by a guanine reduction product. In order to carry out AdSV, accumulation was carried out at either open circuit or between −0.1 and −0.8 V. This was followed by the cathodic scan to a potential of −1.85 V and then the reversed anodic scan to obtain the stripping peak at −0.3 V. This peak increased

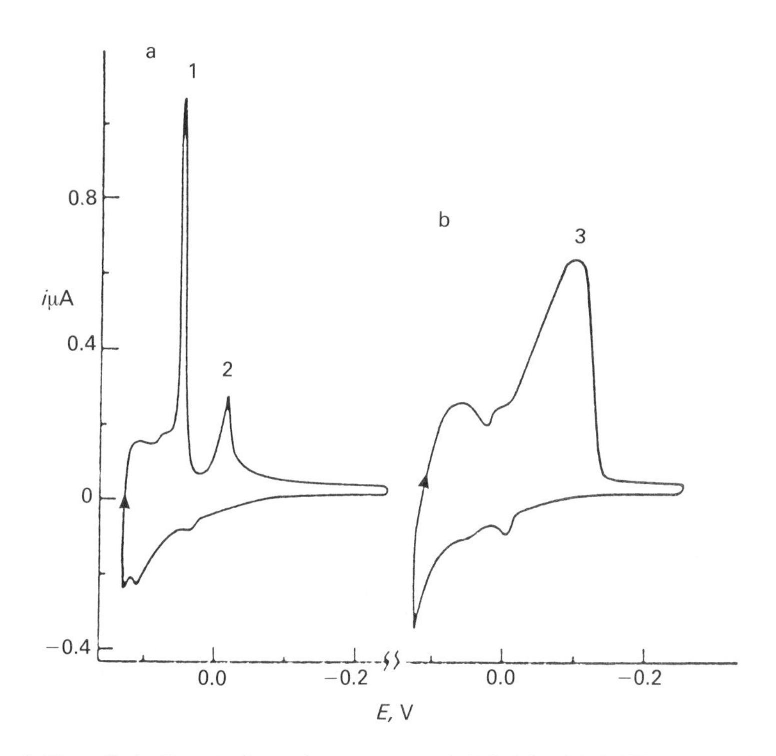

Fig. 2.18 — Cathodic stripping voltammograms of 6 MeAde: (a) 0.057 mM, t_d=20 s: (b) 0.107 mM, t_d=30 s. E_d +125 mV; buffer borax pH 9.2; scan rate 25 mV s^{-1}. (Reprinted with permission from E. Palecek, J. G. Osteryoung and R. A. Osteryoung, *Anal. Chem.*, 1982, 54, 1389. Copyright 1982, American Chemical Society.)

in magnitude with deposition time; however the sensitivity was found to be much greater for denatured DNA than for native DNA (Fig. 2.19). This difference in sensitivity could be exploited for studies involving the thermal degradation of DNA.

2.3.3 Voltammetric methods with carbon electrodes

2.3.3.1 Linear-sweep and cyclic voltammetry

It is well known that LSV can provide rapid, simple and reliable analytical methods; it is not surprising, therefore, that attempts have been made to utilize this technique for determinations of purine- and pyrimidine-containing compounds.

The development of such voltammetric methods for the determination of purine bases and nucleosides, in mixtures containing these two, has been the subject of several investigations; these methods exploit the ability of the compounds to undergo oxidation reactions at carbon electrodes.

In several early studies Dryhurst [60,61] showed that adsorption of adenine, adenosine, guanine and guanosine occurred at the PGE, which affected the voltammetric determinations. However, in the presence of 5- to 10-fold excess of guanosine, adsorbed guanine was displaced and the process became diffusion-controlled,

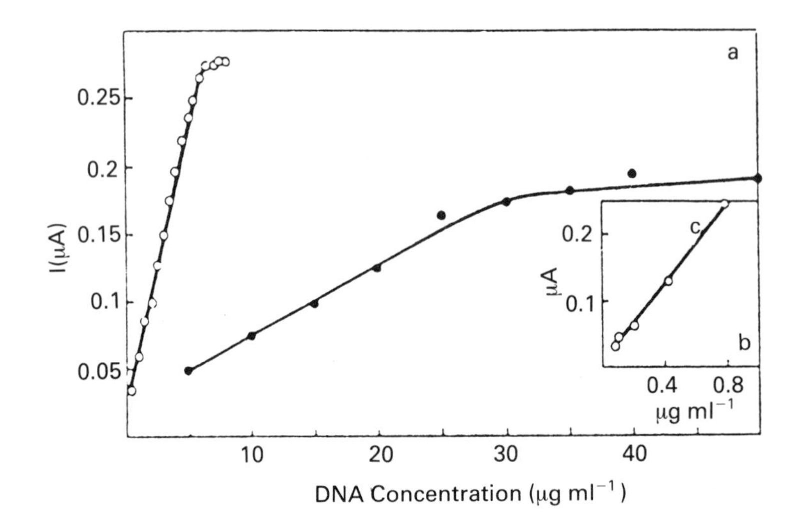

Fig. 2.19 — Dependence of the height of the anodic peak on DNA concentration: (○) denatured DNA; (●) native DNA. (a) t_A=4 min, stirring at 150 rpm, surface area 2.2 mm^2; (b) t_A=8 min, stirring at 200 rpm, surface area 3.5 mm^2. Open circuit accumulation; buffer 0.6 M ammonium formate/0.1 M sodium phosphate pH 6.8; scan rate 200 mV s^{-1}; adsorption potential −0.1 V, accumulation time 140 s. (Reproduced from [59] by permission of the copyright holders, Elsevier Science Publishers Physical Sciences and Engineering Division.)

allowing guanine to be determined without interference; it was also possible to monitor adenine in the presence of adenosine in a similar manner. Unfortunately, the simultaneous measurement of the nucleosides in these studies was difficult owing to the large concentrations required in the test solution. However, in a later study Yao *et al.* [62] showed that the adsorption of adenine and adenosine, as well as guanine and guanosine, could be controlled by the choice of buffer pH. These authors investigated the effect of pH on peak currents and potential over the range 2–12. From the data, derived from plots of current function versus scan rate, they reported that at pH values above 4 reactant was adsorbed on the electrode; but below pH 4 the current was diffusion-controlled, which indicated absence of adsorption. This conclusion was also confirmed by the observation that calibration graphs were curved in pH 7 buffer but linear at pH 3; at this latter pH value linearity was observed down to 0.05 mM. The linear-sweep voltammogram for adenine, adenosine, guanine and guanosine in Britton–Robinson buffer pH 3.0 showed four well-defined peaks suitable for the simultaneous determination of all four substances (Fig. 2.20). It should be added that this could not be achieved by DPP.

Further LSV studies, using a glassy carbon electrode, have been carried out by the same research group on the hydroxypurines hypoxanthine, xanthine and uric acid [63]; the purpose of this investigation was again to develop a method for the simultaneous determination of the three compounds in a mixture. It was shown that adsorption of these substances occurred from buffer solutions at around neutral pH, which was, therefore, unsuitable for quantitative purposes; this was particularly

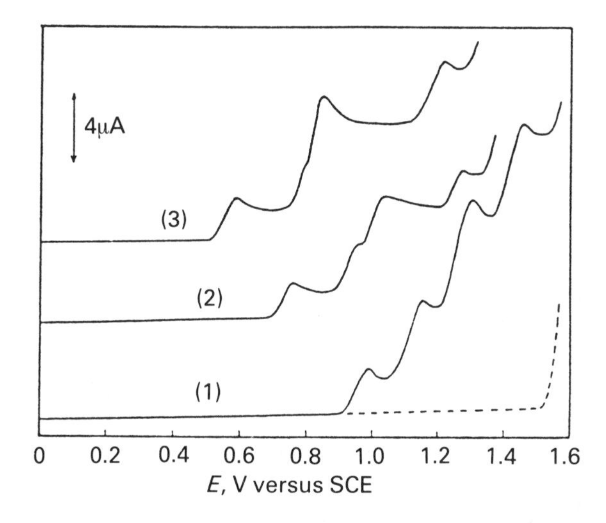

Fig. 2.20 — Anodic voltammograms for the mixtures of adenine, adenosine, guanine and guanosine. Concentration of each of adenine, adenosine and guanosine 0.4 mM, concentration of guanine 0.2 mM. pH (Britton–Robinson buffer); (1) 3.0, (2) 7.0, (3) 10.0. (Reproduced from [62] by permission of the copyright holders.)

problematical with hypoxanthine, which exhibited a marked decrease in current when a 1.0-mM solution was investigated (Fig. 2.21). The most suitable conditions for LSV analysis of a mixture containing the three specified purines were reported to be either 1-m H_2SO_4 or H_3PO_4. Under these conditions the calibration graphs were linear between 0.02 and 0.4 mM.

An interesting approach to the use of CV for the analysis of normal and pathological urine samples was recently described by Moutet *et al.* [64]. The workers used an electrode comprised of graphite powder which was packed in such a way as to allow penetration of the sample into a compartment that was essentially separated from bulk sample. The mass transport through the sinter was slow compared to the rate of electron transfer from the high-surface-area graphite powder electrode to the components. This resulted in very symmetrical triangular peak shapes rather than the usual diffusion-controlled shape. Because of this, the method was readily able to differentiate a number of components in the urine samples; this included a very well defined uric acid peak (Fig. 2.22).

2.3.3.2 Differential-pulse voltammetric methods

A number of purine bases and nucleosides have been submitted to investigations using DPV at a glassy carbon electrode. Kenley *et al.* [65] showed that a number of these were electroactive in 50% methanolic phosphate buffer, pH 7.0, and these included: adenine (E_P 1.04 V); guanine (E_P 0.74 V); 2′-deoxyguanosine (E_P 1.03 V); guanosine (E_P 1.03 V); hypoxanthine (E_P 1.03 V); uric acid (E_P 0.33 V); 2-*N*-acetylguanine (E_P 0.88 V, where potentials are quoted versus Ag/AgCl). The workers described an equation relating the peak potentials to substituent constants; it was suggested that it should be possible to predict the peak potential for other

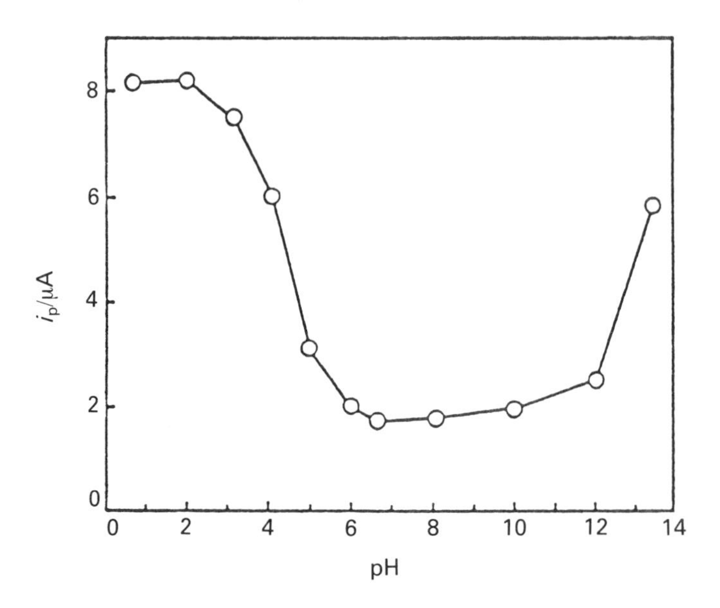

Fig. 2.21 — The effect of pH on the peak current of 1.0 mM hypoxanthine. Glassy carbon electrode geometric area: 7.1 mm^2, scan rate: 3.3 $mV\,s^{-1}$. (Reproduced from [63] by permission of the copyright holders.)

purines when the substitution patterns were known. This data indicated that strong substituent effects could be exploited to determine nucleoside analogue drugs in the presence of their degradation products; such a method had then been developed by the DPV technique [66].

Anodic DPV, with a glassy carbon electrode, was employed by Sequaris *et al.* [67] to study the effect of an alkylating agent on DNA. The effect of this agent was to cause methylation at the 7-position of the guanine molecule, which produced 7-methylguanine. Therefore, the proposed method was developed in order to measure the amount of this substance in nucleic acids. The guanine derivative was found to exhibit an oxidation peak at about +1.25 V in 0.5-M sodium perchlorate buffer that had been adjusted to pH 1.2 with perchloric acid; in the same medium guanine showed a peak at about +1.10 V. In order to eliminate effects from interferences the guanine peak was essentially used as an internal standard and the ratio of 7−Me gua/gua peak currents was plotted versus the ratio of 7-Me gua/gua bulk concentrations. The method for nucleic acid degree of alkylation involved hydrolysis of the nucleic acid with 1-M $HCl0_4$ for 15 min; following dialysis the solution was treated with necessary buffer components and sumitted to DPV. The authors suggested that the procedure provided a convenient and relable approach for *in vitro* screening of the alkylation efficiency of agents exerting mutagenic and carcinogenic effects by their interaction with nucleic acids.

Further investigations which utilized DPV for nucleic acid anlysis were described by Brabec [68]. This author reported on one particular study where the anodic peak currents for adenine and guanine in nucleic acids were used to probe the sites of

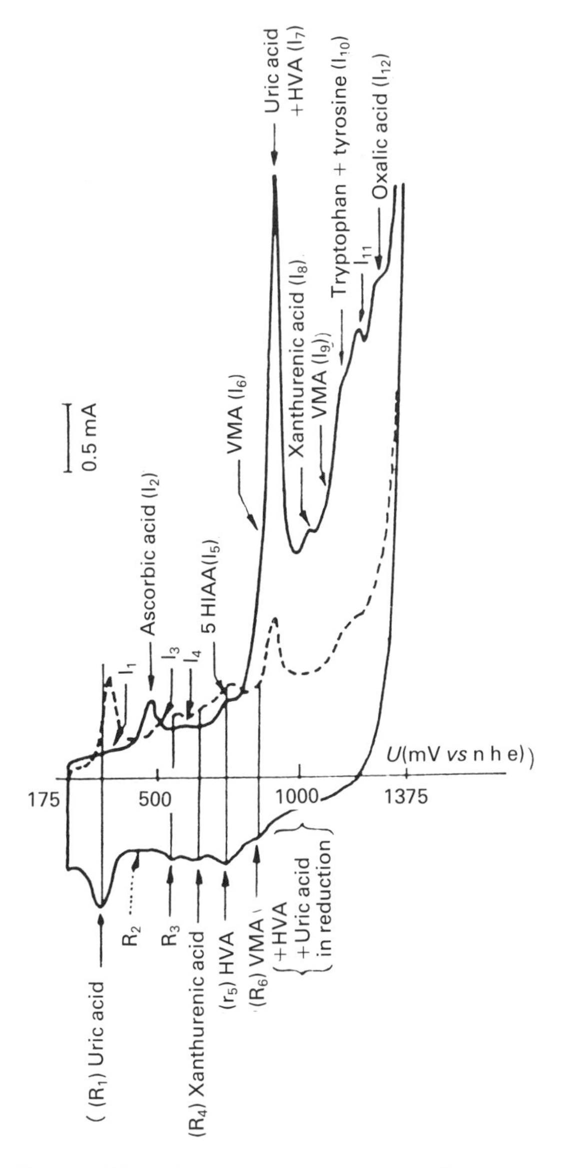

Fig. 2.22 — Identification of the main redox systems present in voltammograms corresponding to a qualitatively normal urine (———) first scan; (- - -) second scan. T 25°C, pH 1.0, scan rate 5 mV s^{-1}. (Reproduced from [64] by permission of the copyright holders, Elsevier Science Publishers.)

denaturation. First of all, it was shown that denatured DNA gave larger peaks for these two bases than the native DNA; this was apparently due to the flexibility of the former, which allowed more bases to contact the rather uneven PGE surface. Next, under conditions of controlled thermal denaturation, a more pronounced increase in the adenine peak was observed; this indicated a more rapid increase in the accessibility of adenine residues as compared to guanine residues for oxidation. Brabec thus concluded that the rate of melting in DNA regions rich in AT pairs is initially higher than that of regions rich in GC pairs. High doses of radiation has similar effects on the voltammetric peaks at the graphite electrode. It was also suggested that the technique could be used to determine the G+C content of DNA. Synthetic polynucleotides were also investigated by the technique but a glassy carbon was used as the working electrode material [69]. It was reported that polyA gave one well-defined peak; the transition of single-stranded polyA to double-stranded polyA, which took place over a narrow pH range, was accompanied by a sharp decrease in the oxidation current (Fig. 2.23). Other polynucleotides have been investigated in a similar manner [70–72].

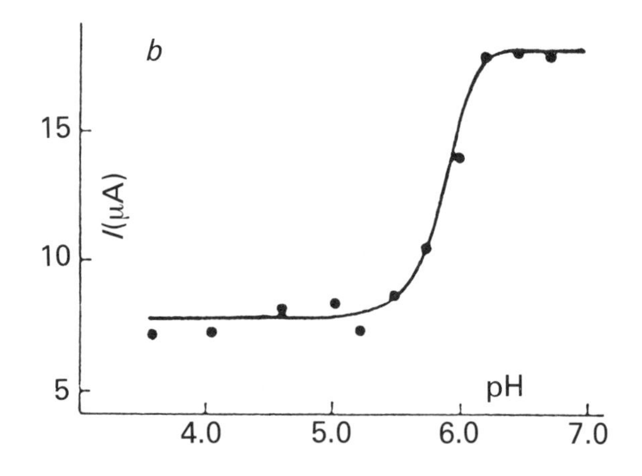

Fig. 2.23 — Variation of the differential pulse voltammetric peak height with pH for 1×10^{-4}-M poly(A) at a glassy carbon electrode in 0.5-M sodium acetate. The poly(A) concentration is related to phosphorus content. (Reproduced from [68] by permission of the copyright holders, Elsevier Science Publishers.)

Several recent reports have indicated that carbon paste electrodes may be suitable for the determination of uric acid in the rat brain [73,74]. Voltammetry using carbon electrodes has also been applied to the determination of purines with pharmacological action.

A novel DPV method, involving *in situ* electrode modification, was developed by Szurley and Brajter-Toth [75] for direct determination of 2,6-diamino purine and its metabolite, 2,6-diamino-8-purinol; the former has been shown to act as a growth inhibitor of bacteria and mammalian cells. These workers showed that when voltammetry was carried out at a PGE in 0.5-M phsophate buffer (pH 7) solutions the resulting calibration plots were found to exhibit curvature; apparently this was a consequence of adsorption at the electrode surface. However, it was discovered that allopurinol was preferentially adsorbed onto the electrode when solutions containing the three compounds were submitted to DPV; the result was that plots for the drug and its metabolite were then linear between 1×10^{-6} and 5×10^{-4} M, which was reported to be suitable for quantitative analysis. It should be mentioned that, bearing in mind the work of Yao *et al.* [62,63], it may have been possible to also circumvent the problem of adsorption by using more acidic buffer solutions.

A method for the determination of theophylline (1,3-dimethylxanthine, a bronchodilator drug) in plasma at therapeutic levels has been described by Munson and Abdine [76]. The method is based on the oxidation of this substance at the stationary carbon paste electrode, the mechanism of which has been described in detail by Hansen and Dryhurst [77]. The DPV method [76] involved adjusting the pH of plasma to a value of 9.0 and extracting with chloroform/2-propanol (95:5, v/v); the separated organic phase was treated with 0.1-M NaOH and an aliquot of the latter was buffered to pH 7.4. The differential-pulse voltammogram was then recorded between +0.5 and +1.2 V versus Ag/AgCl. The only interference found was due to 3-methylxanthine, which appears in blood as a metabolite of theophylline; however, the extraction procedure minimized this interferent. Unfortunately, the authors suggested that their method may not be applicable to patients on long-term theophylline therapy.

2.3.3.3 Stripping voltammetry

As illustrated in a few of the above examples the presence of adsorption in voltammetry can be detrimental to the measurement of the analyte. However, over the past few years this phenomenon has also been exploited to greatly increase both the sensitivity and the selectivity of voltammetric methods. This is possible because adsorption may be used to preconcentrate the compound of interest at the working electrode prior to the voltammetric measurement step [78,79]. Several reports have appeared recently where this approach has been applied to investigations on purine derivatives.

Wang and Freiha [80] have described a method for the determination of uric acid in biological fluids using adsorptive preconcentration at the graphite paste electrode; a differential pulse waveform was used in the measurement step. By simply immersing the electrode in the sample containing uric acid (UA) for a given period of time and then transferring the electrode to an electrolytic blank solution a high degree of selectivity was achieved. Substances that were oxidized at the electrode but not preconcentrated (particularly vitamin C) were simply retained in the sample solution and did not interfere. In this case, it was found that preconcentration proceeded best at low pH (3) and voltammetry at pH 7.4. In order to minimize loss of UA during medium exchange it was reported that flow-injection analysis (FIA) could be used; the sample was injected into a flowing stream of pH 3 to preconcentrate UA and DPV was performed after changing the flowing stream to pH 7.4. Fig. 2.24 shows the voltammograms recorded on standard solutions and dilute urine; as is apparent UA could be detected in the dilute urine sample.

A great enhancement in sensitivity for the voltammetric determination of adenine and guanine at glassy carbon has been achieved by a rather interesting preconcentration technique. This method was reported by Shiraishi and Takahashi [81] and involved pre-electrolysis (at +0.05 V versus SCE) of a solution containing adenine and guanine, in the presence of copper (II) ions (at a concentration of 0.001 M). The resulting DPV peak currents were proportional to concentration in the range 5×10^{-7} to 8×10^{-6} M when the accumulation time was 3 min; the

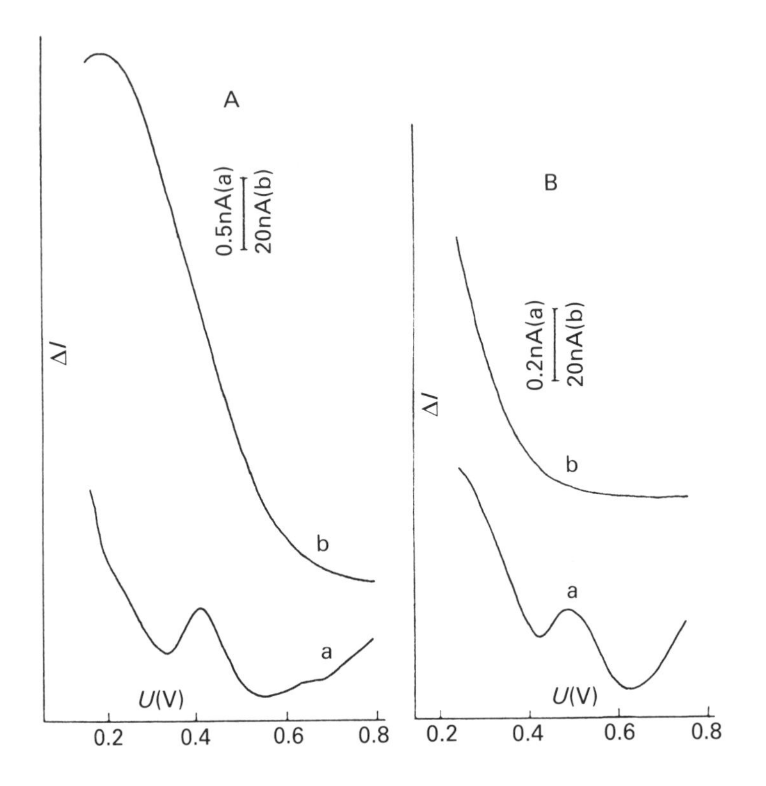

Fig. 2.24 — Differential pulse/flow injection measurements of uric acid. Injected solutions containing 1×10^{-4} M uric acid and 7×10^{-4} M ascorbic acid (A) and diluted (1:9) urine (B). Preconcentration period, (a) 90, (b) 30 s. Scan rate 10 mV s^{-1}; amplitude 50 mV; sample solution 0.2 M KH_2PO_4 (pH 3); exchange solution 0.05-M phosphate buffer pH 7.4. (Reproduced from [80] by permission of the copyright holders, Elsevier Science Publishers.)

supporting electrolyte consisted of 0.4-M acetate buffer pH 4.0. It was stated that the bases could be detected at the 2×10^{-7}-M level by simply using a 5-min accumulation time. The authors considered that the peak enhancement was the result of adsorption of copper (ionic or metallic) compounds of the bases onto the glassy carbon electrode.

2.3.4 Liquid chromatography with electrochemical detection (LCEC)

The oxidation reactions which purines and pyrimidines undergo at solid electrodes (see section 2.2) may be exploited for their determination at low levels (i.e. picomoles per cubic centimetre in some cases) in a variety of mixtures, including complex matrices such as biological fluids, using LCEC. In order to optimize the methods careful consideration has to be given to certain experimental parameters; these include: the configuration of the flow-through cells and mode of operation; mobile phase composition and the column stationary phase. In this subsection examples will be given to illustrate the ways in which LCEC may be optimized and applied to the determination of purine and pyrimidine derivatives.

A number of studies have focussed attention on the application of LCEC for the analysis of UA in biological fluids. In order to determine the optimum applied potential for an amperometric detector, which incorporated a glassy carbon working electrode and which was operated in the oxidative mode, Iwamoto *et al.* [82] constructed a hydrodynamic voltammogram (HDV). Fixed injection volumes (10 μl) containing 1 ng of purine derivative were injected onto a reversed-phase column (Fine Sil, C_{18}, 10 μm; 25 cm×4.6 mm i.d.); the potential was adjusted after each injection until +1,0 V versus Ag/AgCl had been reached (Fig. 2.25). From the HDV a potential of +0.8 V versus Ag/AgCl was selected as the operating potential for the LCEC assay of UA. In this assay the mobile phase consisted of 0.2-M phosphate buffer (KH_2PO_4–H_3PO_4, pH 2.0); the flow rate was 0.5 ml min^{-1}. The method was used successfully for the measurement of UA in rat serum following a protein precipitation stage with sulphosalicylic acid (SSA); the concentration was reported to be 4.9 ng/ml and the minimum detectable quantity was 10 pg.

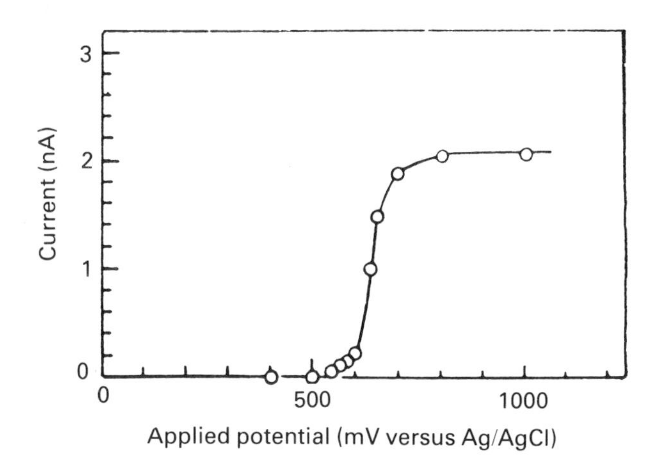

Fig. 2.25 — Hydrodynamic voltammogram for uric acid obtained by repeated injection of 1 ng of uric acid at different electrochemical detector potentials. (Reproduced from [82] by permission of the copyright holders, Elsevier Science Publishers Physical Sciences and Engineering Division.)

Further investigations on UA were carried out by the same group of workers [83,84] for the determination of this substance in human serum. Human serum (0.5 ml) was mixed with 1.5 ml of an aqueous solution containing 2% *m*-phosphoric acid and the mixture was centrifuged for 30 min. The supernatant was passed through a membrane filter to remove particulate matter; then 10 μl of the filtrate was injected onto the column (5 cm×4.6 mm i.d.). Since the column was shorter, and the flow rate had been increased to 2 ml min^{-1} in this study, complete separation of UA in serum was achieved in only 2 min. Although concentrations of UA and ascorbic acid (AA) in human serum were found to be 10.6 and 93.3 μg cm^{-3} respectively, the latter peak was not well resolved from the solvent peak.

The same research group have continued their studies on the anlysis of UA levels, in the presence of AA, in a variety of human biological fluids; they were successful in achieving well-resolved peaks for both compounds by the use of a new rigid-type porous polymer as the packing material (polymetacrylate gel; RSpak DE-13, Showadenko K. K., Tokyo) [85,86]. This was used in conjunction with a mobile phase containing 0.1-M dipotassium hydrogen phosphate–citric acid buffer, adjusted to pH 4.6, containing 1-mM EDTA (ethylenediamide tetraacetic acid). Electrochemical detection was achieved as described above; Fig. 2.26, shows the chromatograms obtained for urine, plasma and cerebrospinal fluid.

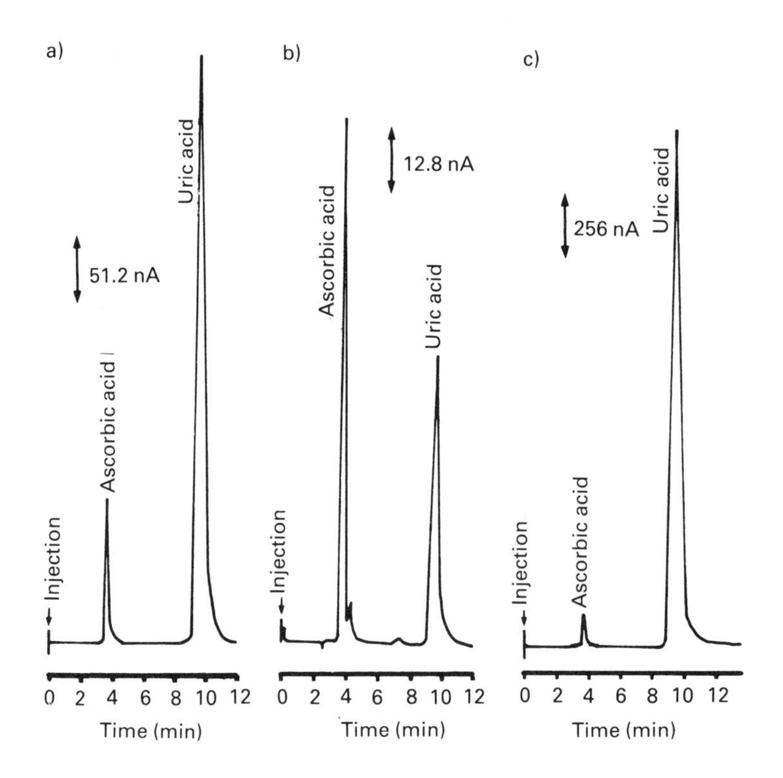

Fig. 2.26 — Typical chromatograms of the simply deproteinized human (a) serum, (b) cerebrospinal fluid, and (c) urine. An aliquot (10 μl) of each simply deproteinized body fluid was injected onto the porous polymer column. (Reprinted from [85] by courtesy of Marcel Dekker Inc.)

Within the last five years several groups have attempted to use various LCEC methods to separate and quantitate mixtures of different complexity containing a variety of purine and pyrimidine derivatives.

A detailed systematic study was recently carried out by Henderson and Griffin [87] to determine the optimum LCEC conditions for the determination of purine bases and nucleosides. In this study a glassy carbon working electrode was incorporated into a thin-layer cell which was operated in the oxidative mode. The column was an ODS Altex Ultrasphere reversed-phase column. With increasing methanol concentration in the mobile phase the electrochemical detector response for hypoxanthine (HX), guanosine GUO, adenosine (ADO), and deoxyadenosine (dADO) decreased considerably; however, the inosine and adenine response decreased only slightly (Fig. 2.27). Detector response (nanoamps per picomole injected) for HX, inosine and ADO was found to increase as a function of phosphate (electrolyte) concentration of the mobile phase up to about 0.04-M potassium phosphate and remained constant between 0.04 and 0.07 M; from the data a concentration of 0.05-M phosphate was selected as being suitable for these substances. The effect of pH was also investigated over the range 3.8–6.8 for five of these compounds; inosine showed an increase in signal magnitude with increasing pH whereas the other four substances showed a decrease (Fig. 2.28). In the same study the authors compared the electrochemical detector with a UV detector, under the same conditions; the results obtained are shown in Table 2.4. As is evident, LCEC appeared to give better sensitivity for purine derivatives, whereas UV detection was better for the pyrimidine compounds.

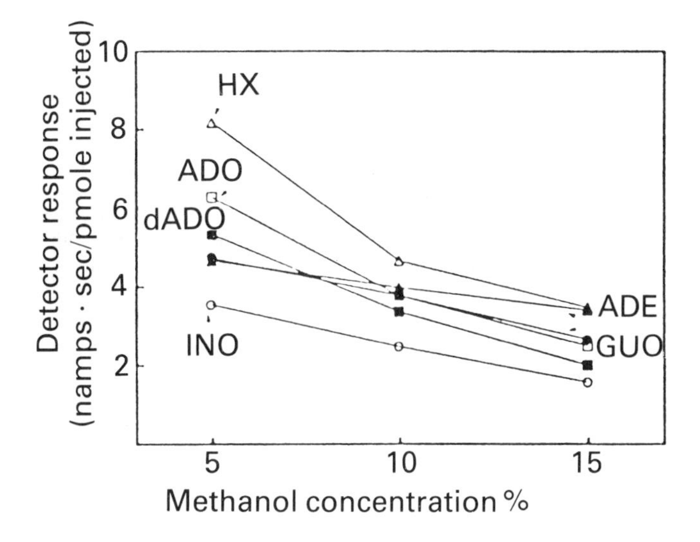

Fig. 2.27 — Effect of methanol content of the chromatographic buffer on the HPLC electrochemical detector response (peak area per picomole injected) for hypoxanthine (HX, 50 pmol), adenosine (ADO, 100 pmol), deoxyadenosine (dADO, 200 pmol), guanosine (GUO, 50 pmol), adenine (ADE, 50 pmol) and inosine (INO, 100 pmoles). Electrode potential +1.5 V; electrochemical detector sensitivity, 100 n A V^{-1}; injection volume, 10 μl; chromatographic buffers 0.05-M potassium phsophate (pH 5.5)-methanol (% v/v as shown). (Reproduced from [87] by permission of the copyright holders, Elsevier Science Publishers Physical Sciences and Engineering Division.)

Further studies on mixtures containing purines have been carried out by Dwyer and Brown [88]; in this investigation a Partisil 10 ODS-3 column and a detector incorporating a glassy carbon working electrode was employed. In this case, the

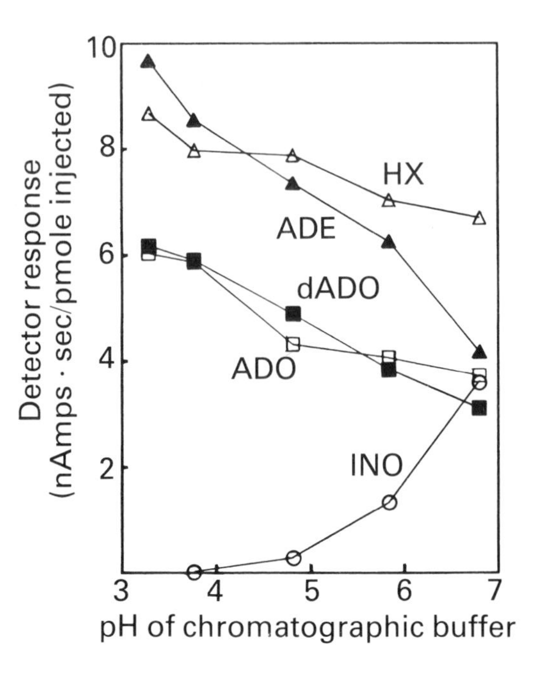

Fig. 2.28 — Effect of chromatographic buffer pH on HPLC electrochemical detector response (peak area per picomole injected) for: HX 40 pmol, ADE 50 pmol, dADO 200 pmol, ADO 100 pmol, and INO 200 pmol. Electrode potential 1.5 V; electrochemical detector sensitivity 100 nA V^{-1}; injection volume 10 μl. Five buffers at different pH values were prepared from 0.05-M potassium phosphate (pH 3.0, 3.5, 4.5, 5.5, 6.5,)–methanol (85:15, v/v) resulting in phosphate methanol chromatographic buffers of 0.0425-M phosphate and pH values of 3.28, 3.76, 4.81, 5.84 and 6.8 respectively. (Reproduced from [87] by permission of the copyright holders Elsevier Science Publishers Physical Sciences and Engineering Division.)

Table 2.4 — Lower-limit of detection using a UV spectrophotometric detector — comparison of electrochemical and UV detection sensitivity

Compound	Amount (pmol)	Volume injected (μl)	Peak height (mm) (mean±S.D)	Relative sensitivity* ED/UV
Adenosine	2.0	20	18.5±0.3	10
Deoxyadenosine	3.0	15	23.2±1.4	6
Adenine	2.5	25	19.0±1.0	50
Inosine	1.5	15	23.5±0.5	*ca.* 1
Hypoxanthine	2.0	20	22.0±1.0	4
Guanosine	3.0	30	22.5±1.2	24
Thymidine	2.5	5	24.2±1.5	*ca.* 1/8
Cytidine	2.5	5	28.1±1.0	*ca.* 1/4

*Relative sentivity, ED/UV $= \frac{\text{minimum detectable amount, electrochemical detector}}{\text{minimum detectable amount, UV detector}}$

(Reproduced from [87] by permission of the copyright holders, Elsevier Science Publishers Physical Sciences and Engineering Division.)

compounds studied were: UA, xanthine, HX, guanosine, xanthosine 5′-phosphate, and guanosine 5′-phosphate, as well as the two important amino acids tyrosine and L-tryptophan. The mobile phase contained 0.02-M potassium dihydrogen phosphate–3% methanol (v/v) adjusted to pH 5.7 with dilute potassium hydroxide solution; the flow rate was 1 ml min^{-1}. In the study it was shown that judicious selection of the operating potential was crucial to the elimination of interferences; above +0.9 V versus Ag/AgCl selectivity became inadequate to resolve all the components of the mixture. Table 2.5 gives the appropriate potentials reported by the authors together with the detection limits for the various compounds.

Table 2.5 — Applied potentials and detection limits of standards

Compound	Potential (V)	Detection limit (pg/ml)
Uric acid	+0.50	16
Tyrosine	+0.75	22
Guanosine	+0.95	250
Xanthine	+0.90	275
L-Tryptophan	+0.80	24
Hypoxanthine	+0.95	334
Guanosine 5′-phosphate	+0.75	42
Xanthosine 5′-phosphate	+0.95	374

(Reproduced from [88] by permission of the copyright holders, Elsevier Science Publishers Physical Sciences and Engineering Division.)

Ghe *et al.* [89] have compared reversed-phase and anion exchange columns and suggested that the latter was more suitable in an LCEC method for the simultaneous determination of eight electroactive purine derivatives. Fig. 2.29 shows the chromatogram obtained with the SAX column and a mobile consisting of potassium dihydrogen phosphate (0.01 M) and acetonitrile (90/10 v/v) adjusted to pH 3.3 with 0.05 M sulphuric acid. The authors reported that the limits of detection were: UA 10 pmol, xanthine pmol, guanine 1 pmol, guanosine 2 pmol, xanthosine 2 pmol, adenine 2 pmol, 6-*N*-dimethylaminopurine 10 pmol.

Several recent reports have been concerned with the development of suitable LCEC procedures for the determination of structurally similar guanine derivatives in rat tissues.

A method was described by Perret [90] which incorporated a glassy carbon electrode in the wall-jet configuration (set at +1.15 V versus Ag/AgCl); this was used to monitor the eluate from a reversed-phase column that contained guanosine, GMP, GDP and GTP from rat kidney. The column was 3-μm ODS Hypersil and the mobile phase contained phosphate buffer pH 7 and an ion pairing agent (tetrabutylammonium chloride); the methanol content was 2% v/v. In this procedure the complete chromatogram was recorded in under 15 min. In order to improve selectivity for several guanine nucleotides a dual-electrode electrochemical detection system has been employed for their determination in the rat brain [91]. In this investigation the glassy carbon working electrodes were arranged in the parallel configuration with one electrode held at +0.95 V and the other held at +0.70 V

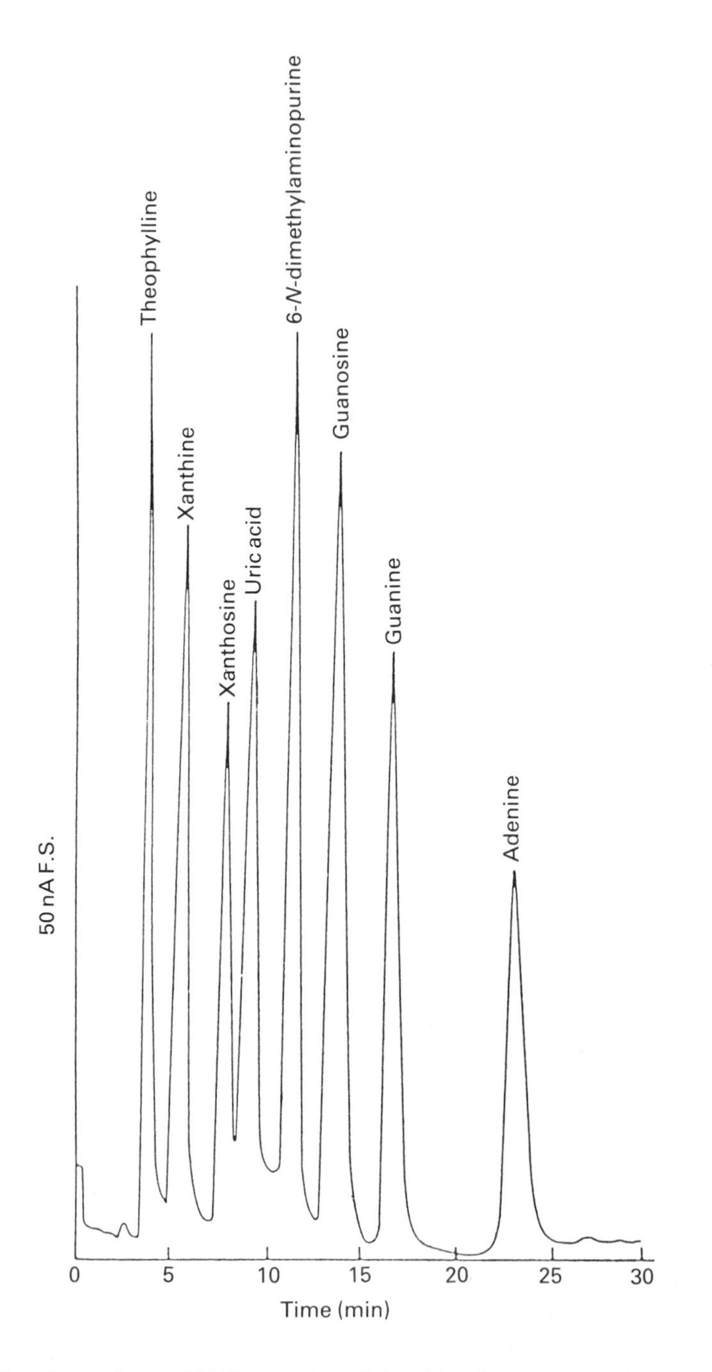

Fig. 2.29 — Ion-exchange HPLC separation of the eight electroactive purine substances examined on the SAX 10 column; 20 μl of solution corresponding to 0.2 nmol of each solute). electrochemical detector +1.20 V versus Ag/AgCl. (Reprinted with permission from *Talanta*, Vol. 33, Page 379, A. M. Ghe, G. Chiavari and J. Evgenidis, 1986. Copyright 1986, Pergamon Press PLC.)

versus Ag/AgCl. Under the conditions used, guanine nucleotides, but not adenine and cytosine nucleotides, were reported to be electroactive at +0.95 V; at 0.7 V guanine nucleotides were electrochemically inactive. However, possible interfering compounds, such as catecholamines and indolamines, were electroactive at both potentials so that their presence in samples could be clearly ascertained. Therefore, selectivity for the the guanine nucleotides in rat tissue was reported to be improved by the proposed method (Fig. 2.30). The chromatographic peak heights for GTP, GDP, GMP and cyclic GMP were linear between 0.5 pmol and 1.0 nmol, the mobile phase consisted of 0.05-M sodium phosphate buffer pH 7.0 with 2.5-mM TBA-methanol (92:8 v/v). The regional distribution of these substances in the rat brain was determined using the developed LCEC assay.

Growing interest in recombinant DNA technology has recently resulted in the development of an LCEC method for the quantitation of nucleic acids at the picogram level [92]. The method was based on hydrolysis and quantitation of the purine bases adenine and guanine. This was regarded as an advantage because pure well-characterized samples for standards could be readily obtained unlike fluorescence methods which rely on DNA standards. The procedure involved the use of a microbondapak C-18 reversed-phase column and a mobile phase consisting of phosphate buffer, pH 7.0, containing 10% methanol; the amperometric detector contained a glassy carbon working electrode held at +1.1 V. The samples investigated included DNA from: calf thymus, salmon testes, *E. coli,* and human placenta.

Methods incorporating HPLC with electrochemical detection have been described for the determination of exogenous purine and pyrimidine compounds in biological fluids; these possessed pharmacological activity.

Smyth [93] reported a method for the diuretic azidopyrimidine (2-amino-4-azido-5-(2-ethoxyethyl)-6-phenylpyrimidine: SC 16102) in urine. For this assay, an aliquot of urine was extracted with dichloromethane and the organic phase was evaporated to dryness; the residue was dissolved in 100 μl of eluting solvent and 20 μl was injected onto a RP-8 column. The eluting solvent was acetonitrile–water, containing 0.05-M lithium perchlorate and the flow-rate was 0.8 ml min; the glassy carbon working electrode was operated at a potential of +1.4 V versus Ag/AgCl. The calibration graph was found to be linear in the range 3–1000 ng.

The determination of selected methyxanthines in serum using LCEC was described by Lewis and Johnson [94]. These authors examined the relative merits of d.c., pulse and differential-pulse amperometry with a cell containing a tubular glassy carbon electrode; the pulse amperometric mode was chosen because it was found to possess the greatest sensitivity. The method was applied to the determination of theophylline (1,3-dimethyxanthine) in serum and the pharmacokinetic response was found following a 300-mg oral dose. The authors reported that this method compared favourably with an HLPC method incorporating a UV detector.

An LCEC method was reported for the determination of thiopurines in plasma in which the detector contained a cobalt-phthalocyanine chemically modified electrode [95]. The advantage of this over an unmodified carbon paste electrode is that a lower working electrode potential can be used; this results in greater selectivity and sensitivity. This method was capable of detecting concentrations as low as 1.5–8.0 pmol, which was approximately an order of magnitude better than that obtained with UV or fluorescence detection.

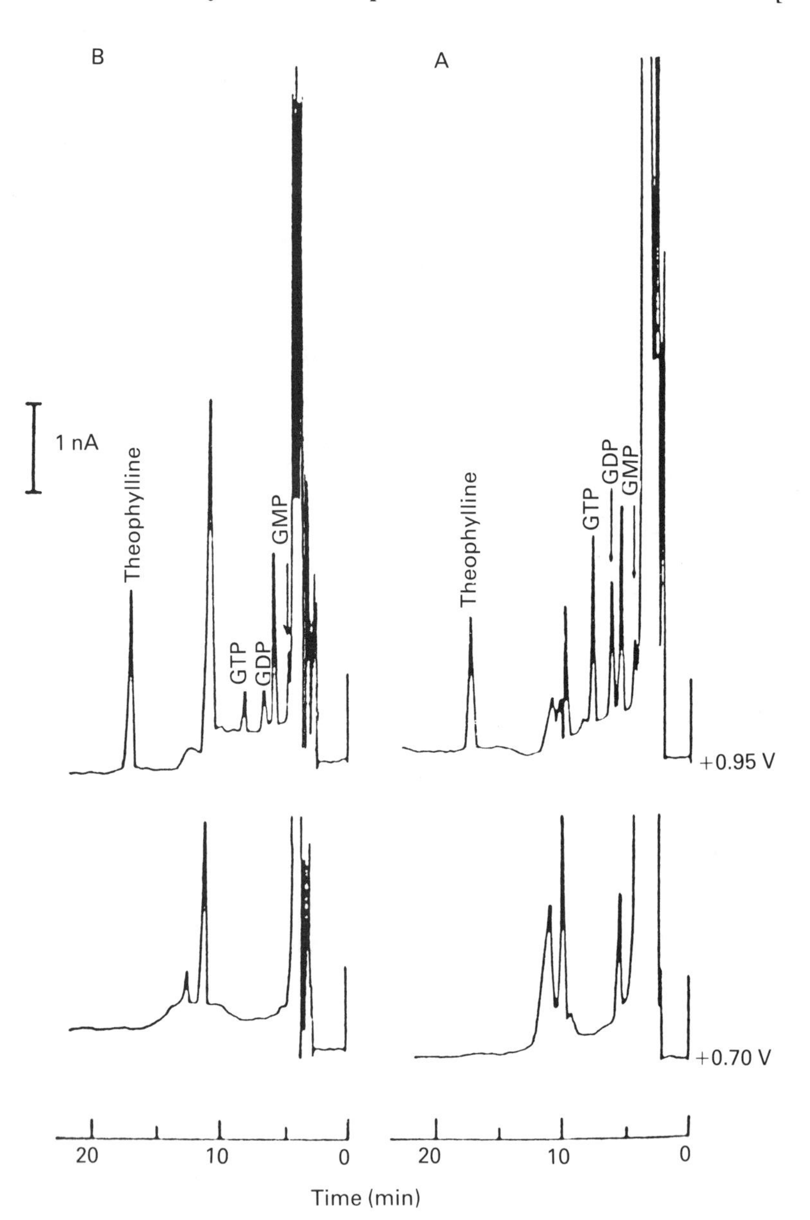

Fig. 2.30 — Chromatograms of brain tissues: (A) cerebral cortex; (B) hypothalmus. Mobile phase 0.05-M sodium phosphate buffer pH 7.0 containing 2.5-mM TBA–methanol (92:8 v/v); flow rate 1.0 ml min^{-1}; GMP, guanosine monophosphate; GDP, guanosine diphsophate; GTP, guanosine triphosphate; theophylline (internal standard). (Reproduced from [91] by permission of the copyright holders Academic Press Inc.)

2.3.5 Methods involving amperometric sensors and biosensors

Over the past few years there has been interest in the application of amperometric sensors for the determination of purine derivatives; such methods have generally been developed to enhance selectivity while maintaining simple experimental set-ups and procedures.

Mckenna and Brajter-Toth [96] have constructed a xanthine oxidase electrode with the conducting salt tetrathiafulvalene tetracyanoquinodimethane, TTF^+TCNQ^- as the electrode material; this was evaluated for the determination of purine, HX and xanthine. The enzyme was immobilized by trapping a drop ($0.05\ cm^3$) of solution containing 90-μM enzyme between the electrode and a membrane prepared from dialysis tubing. In addition to the enzyme working electrode the cell also contained a platinum foil counter electrode and an SCE reference electrode. All current measurements were made at a constant potential of 225 mV versus SCE following a 30-s incubation period at open circuit in the solution containing the analyte; the supporting electrolyte consisted of 0.05-M phosphate buffer pH 8.0. The authors reported that the response of the electrode for purine depended on two reactions:

$$\text{purine}+\text{xanthine oxidase/FAD}\rightarrow\text{uric acid}+\text{xanthine oxidase/FADH}_2 \qquad (2.1)$$

$$\text{xanthine oxidase/FADH}_2\rightarrow\text{xanthine oxidase/FAD}+2e^-+2H^+ \qquad (2.2)$$

where reaction (2.2) occurred at the electrode surface. Both xanthine and HX give similar reactions to (2.1) and (2.2), producing the reduced form of the enzyme and then elctrochemical reoxidation. However, greatest sensitivity was found for purine because the enzyme converts this compound first to hypoxanthine then to xanthine; therefore it effectively turns over three times. The range of linearity for purine and HX was 1.6×10^{-5} M to 6.4×10^{-4} M and for xanthine 1.2×10^{-5} to 1.7×10^{-3} M. The electrode was applied to the determination of hypoxanthine and purine in plasma; although the response decreased compared with the response obtained in buffer it was still possible to carry out analysis in this matrix. Unfortunately, the authors did not include any results for the circulating levels of these two compounds in the plasma taken from normal subjects.

Several reports have appeared that describe the development of amperometric enzyme sensors for the simultaneous determination of purine compounds. In one study [97], the compounds investigated were HX, inosine (HXR), inosine-5′-phosphate (IMP) and adenosine-5′-phosphate (AMP); the simultaneous determination of these four compounds required the use of four enzyme sensors in a flow-through cell. These compounds were determined from oxygen consumption according to the reactions

$$\text{AMP}\xrightarrow{\text{AD}}\text{IMP}\xrightarrow{\text{NT}}\text{HXR}\xrightarrow{\text{NP,PO}_4^{3-}}\text{HX}\xrightarrow{\text{XO,O}_2}\text{UA} \qquad (2.3)$$

where AD is AMP deaminase, NT is nucleotidase, NP is nucleotide phosphorylase and XO is xanthine oxidase. Enzymes were covalently bound to a membrane prepared from cellulose triacetate, 1,8-diamino-4-aminomethyloctane and glutaraldehyde. A 0.05-M phosphate buffer solution (pH 7.8) containing cysteine (10^{-4} M) was continuously transferred to the cell by a peristaltic pump. When the output currents of all electrodes had become steady a known volume (20–50 μl) of sample solution was injected into the flowing buffer solution. The output currents of the oxygen electrodes were recorded simultaneoulsy on a multichannel recorder. The concentration of the purines was then found from the response measured as the difference between the initial current and the minimum current. The multielectrode enzyme sensor was applied to the determination of the four purine derivatives in a number of fish and shellfish; determination of the compounds could be made within 5 min. It was stated that for the shellfish investigated in this study, AMP was decomposed rapidly to ADO; therefore, the decrease in AMP concentration could be used for freshness evaluation. The authors indicated that this sensor was superior to previous sensors [98–100] because the latter required several steps in the determination and were rather time-consuming.

Watanabe *et al.* [101] have improved the design of their sensor so that both HXR and HX could be determined separately with only one electrode. The device consisted of membranes prepared from cellulose triacetate and 1,8-diamino-4-aminomethyloctane; the xanthine oxidase and nucleoside phosphorylase enzymes were covalently immobilized onto separate sheets of this. The xanthine oxidase membrane was followed by three sheets of the untreated membrane and finally the membrane containing nucleoside phosphorylase was placed on the tip of the oxygen electrode (Fig. 2.31). The mechanism for the simultaneous determination HX and

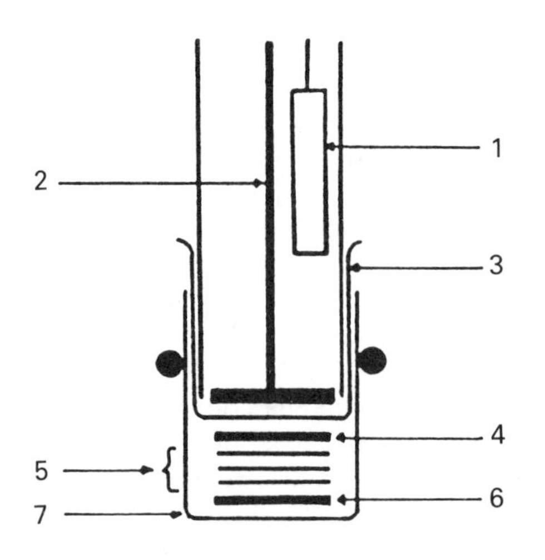

Fig. 2.31 — Schematic diagram of the tip of the enzyme electrode. 1, Pb anode; 2, Pt cathode; 3, Teflon membrane; 4, nucleoside phosphorylase membrane; 5, triacetyl cellulose membranes containing 1,8-diamino-4-aminomethyloctane; 6, xanthine oxidase; 7, dialysis membrane. (Reproduced from [101] by permission of the copyright holders Elsevier Applied Science Publishers.)

HXR is illustrated in Fig. 2.32. The oxidation of HX to UA by xanthine oxidase causes oxygen consumption; a decrease in dissolved oxygen around the membrane results and the current from the electrode decreases. Since HXR does not react with xanthine oxidase oxygen decrease does not occur. However, HXR does react at the second enzyme membrane to produce HX; when this diffuses to the xanthine oxidase electrode a further decrease in current occurs. Therefore, simultaneous determi-

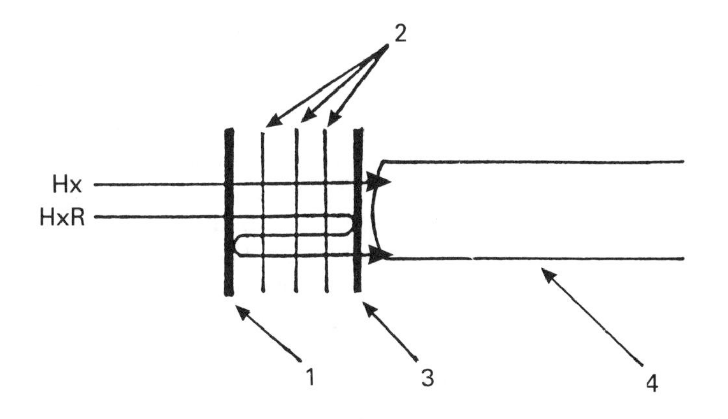

Fig. 2.32 — Separation mechanism of hypoxanthine and inosine, 1, xanthine oxidase membrane; 2, triacetyl cellulose membranes containing 1,8-diamino-4- aminomethyloctane; 3, nucleoside phosphorylase membrane; 4, oxygen electrode; HX, hypoxanthine; HXR, inosine. (Reproduced from [101] by permission of the copyright holders Elsevier Applied Science Publishers.)

nation of the two purines is possible by using the lag period for the intiation of the xanthine oxidase reaction. The conditions for this sensor have been optimized; linear relationships between current and concentration were obtained in the range 0.1–0.4 mM and 1.0–4.0 mM for HX and HXR respectively.

2.4 CONCLUSION

The first part of this chapter discussed the oxidation/reduction processes which occurred at mercury and carbon electrodes for some representative purine and pyrimidine derivatives. In some cases, it was suggested by the research workers that the electrode mechanisms may mimic enzyme reactions; therefore, electrochemistry may be of value in the study of enzyme reactions in biological systems.

The second part of the chapter described the exploitation of the electrochemical behaviour of these derivatives for their determination in a variety of samples. These methods involved the application of various modern electroanalytical techniques; it was the intention to consider the ways in which these were used to overcome analytical problems in selected areas.

One area of particular interest has been nucleic acid analysis and most of the modern voltammetric techniques (involving both mercury and carbon electrodes) have found application in this area. Important investigations have been carried out to study the effects of radiation and chemical treatment on double-helical DNA *in vitro*. It has been possible to readily detect changes in the conformation of native DNA after such treatment and it seems likely that these techniques could be used to test for potential carcinogenic materials by *in vitro* methods; therefore, in the future it may be possible to eliminate the use of live animals in some types of investigation.

Another area of interest has been in the use of electrochemical sensors, based on the amperometric detection of purine-containing metabolites, for the determination of food freshness. Although in this review only fish products have been mentioned it seems feasible that other food types could also be monitored by such techniques.

Electroanalytical techniques have also been used in studies on body fluids and tissues, of human and animal origin. For example, the application of voltammetric techniques, using carbon electrodes, has resulted in rairly rapid, direct and reliable methods for the determination of UA in human urine samples. LCEC has been successfully applied to the simultaneous determination of a variety of naturally occurring nucleosides and nucleotides in body fluids. These are clearly very difficult analyses since the species are usually present at very low concentrations and they are structurally very similar; in addition, interference may be also occur from other naturally occurring substances. Therefore, the advantages of this technique in such determinations are obvious. It should be added that there are many other possible derivatives which might be amenable to this type of approach.

The determination of purine and pyrimidine derivatives, which act as drug substances, has also been described using voltammetric and LCEC methods; because the latter are more selective and sensitive they would be more suitable for the determination of blood levels in the nanomolar (and lower) range. It seems likely that other drugs of this type will emerge in the future and, no doubt, will be amenable to analysis using advanced electroanalytical techniques.

REFERENCES

[1] E. Palecek, *Bioelectrochem. Bioenerg.*, 1981, **8**, 469.
[2] E. Palecek, in *Topics in Bioelectrochemistry and Bioenergetics*, G. Milazzo (Ed.), pp 65–157, John Wiley and Sons, 1983.
[3] K. Diem and C. Lentner, (Eds), *Documenta Geigy Scientific tables*, Geigy Pharmaceuticals, Macclesfield, 1985, p 351.
[4] B. Janik and P. J. Elving, *Chem. Rev.*, 1968, **68**, 295.
[5] K. Diem and C. Lentner, (Eds), *Documenta Geigy Scientific tables*, Geigy Pharmaceuticals, Macclesfield, 1985, p 354.
[6] W. F. Hemmens, *Biochim. Biophys. Acta*, 1964, **91**, 332.
[7] D. B. Dunn, *Biochem. J.*, 1963, **86**, 14.
[8] K. Diem and C. Lentner, (Eds), *Documenta Geigy Scientific tables*, Geigy Pharmaceuticals, Macclesfield, 1985, p 671.
[9] British National Formulary, British Medical Association and the Pharmaceutical Society of Great Britain, No. 7, 1984, p. 106.
[10] J. P. Hart, in *Electrochemistry Sensors and Analysis*, M. R. Smyth and J. G. Vos (Eds.) p. 355, Elsevier, Amsterdam, 1986.
[11] P. T. Kissinger and W. R. Heineman (Eds) *Laboratory Techniques in Electroanalytical Chemistry*, Marcel Dekker, 1984.
[12] W. F. Smyth, (Ed.), *Polarography of Molecules of Biological Significance*, Academic Press, London, 1979.
[13] A. M. Bond, *Modern Polarographic Techniques in Analytical Chemistry*, Marcel Dekker, New York, 1980.
[14] P. J. Elving, S. J. Pace and J. E. O'Reilly, *J. Am. Chem. Soc.*, 1973, **95**, 647.
[15] G. Dryhurst and P. J. Elving, *Talanta*, 1969, **16**, 855.
[16] B. Janik and E. Palecek, *Arch. Biochem, Biophys.*, 1964, **105**, 225.
[17] E. Palecek and B. Janik, *Arch. Biochem. Biophys.*, 1962, **98**, 527.
[18] O. Manousek and P. Zuman, *Chem. Listy*, 1956, **49**, 668.
[19] O. Manousek and P. Zuman, *Collect. Czech. Chem. Commun.*, 1955, **20**, 1340.
[20] V. Brabec, *Bioelectrochem. Bioenerg.*, 1981, **8**, 437.
[21] D. L. Smith and P. J. Elving, *J. Am. Chem. Soc.*, 1962, **84**, 1412.
[22] E. Palecek, F. Jelen, M. A. Hung and J. Lasovsky, *Bioelectrochem. Bioenerg.*, 1981, **8**, 621.
[23] E. Palecek, J. Osteryoung and R. A. Osteryoung, *Anal. Chem.*, 1982, **54**, 1389.
[24] G. Dryhurst and G. F. Pace, *J. Electrochem. Soc.*, 1970, **117**, 1259.

[25] R. N. Goyal and G. Dryhurst, *J. Electroanal. Chem.*, 1982, **135**, 75.
[26] G. Dryhurst, *Electrochemistry of Biological Molecules,* Academic Press, 1977.
[27] A. C. Conway, R. N. Goyal and G. Dryhurst, *J. Electroanal. Chem.*, 1981, **123**, 243.
[28] G. Dryhurst, N. T. Nguyen, M. Z. Wrona, R. N. Goyal, A. Brajter-Toth, J. L. Owens, and H. A. Marsh, *J. Chem. Ed.*, 1983, **60**, 315.
[29] R. N. Goyal, A. Bratjer-Toth and G. Dryhurst, *J. Electroanal. Chem.*, 1982, **131**, 181.
[30] D. L. Smith and P. J. Elving, *Anal. Chem.*, 1962, **34**, 930.
[31] T. E. Cummings, J. R. Fraser and P. J. Elving, *Anal. Chem.*, 1980, **52**, 558.
[32] Y. M. Temerk, M. M. Kamal, M. E. Ahmed and M. Ibrahim, *Bioelectrochem. Bioenerg.*, 1983, **11**, 449.
[33] Y. M. Temerk, Z. A. Ahmed, M. E. Ahmed and M. M. Kamal, in *Electrochemistry, Sensors and Analysis,* M. R. Smyth and J. G. Vos (Eds), p. 229, Elsevier, Amsterdam, 1986.
[34] Y. Kato and A. Hatano, *Analyst,* 1983, **108**, 122.
[35] B. Bouzid and A. M. G. Macdonald, *Anal. Chim. Acta,* 1988, **211**, 155.
[36] E. Palecek, *Electroanalysis in Hygiene, Environmental, Clinical and Pharmaceutical Chemistry,* W. F. Smyth (Ed.), p. 79, Elsevier, 1980.
[37] V. Brabec and E. Palecek, *Biopolymers,* 1972, **11**, 2577.
[38] E. Palecek, in *Progress in Nucleic Acid Research and Molecular Biology,* W. E. Cohn (Ed.), vol. 18, p. 151, Academic Press, New York, 1976.
[39] V. Brabec and E. Palecek, *Z. Naturforsch,* 1973, **28**, 685.
[40] E. Palecek and J. Doskocil, *Anal. Biochem.*, 1974, **60**, 518.
[41] M. Vorlickova and E. Palecek, *Biochim. Biophys. Acta,* 1978, **517**, 308.
[42] J.-M. Sequaris and P. Valenta, in *Electroanalysis in Hygiene, Environmental, Clinical and Pharmaceutical Chemistry,* W. F. Smyth (Ed.), p. 99, Elsevier, 1980.
[43] J.-M. Sequaris, in *Electrochemistry, Sensors and Analysis,* M. R. Smyth and J. G. Vos (Eds), p. 191, Elsevier, Amsterdam, 1986.
[44] E. Palecek, *Collect. Czech. Chem. Commun.*, 1974, **39**, 3449.
[45] V. Brabec and E. Palecek, *Biophys. Chem.*, 1976, **4**, 79.
[46] E. Palecek and V. Vetrl, *Bioelectrochem. Bioenerget.*, 1977, **4**, 361.
[47] J.-M. Sequaris, H. W. Nurnberg and P. Valenta, *Toxicol. Environm. Chem.*, 1985, **10**, 83.
[48] E. Palecek, F. Jelen and L. Trnkova, *Gen. Physiol. Biophys.*, 1986, **5**, 315.
[49] B. Bouzid and A. M. G. Macdonald, *Anal. Chim. Acta,* 1988, **211**, 175.
[50] E. Palecek, *Anal. Biochem.*, 1980, **108**, 129.
[51] E. Palecek and F. Jelen, *Collect. Czech. Chem. Commun.*, 1980, **45**, 3472.
[52] T. M. Florence, *J. Electroanal. Chem.*, 1979, **97**, 219.
[53] E. Palecek, *Anal. Chim. Acta,* 1985, **174**, 103.
[54] B. Bouzid and A. M. G. Macdonald, *Anal. Proc.*, 1986, **23**, 295.
[55] A. J. Miranda Ordieres, M. J. Garcia Gutierrez, A. Costa Garcia, P. Tunon Blanco, W. F. Smyth, Analyst, 1987, **112**, 243.
[56] J. Wang, M. S. Lin and V. Villa, *Analyst,* 1987, **112**, 247.
[57] B. Bouzid, PhD Thesis, University of Birmingham, 1987.
[58] E. Palecek and Mac Anh Hung, *Anal. Biochem.*, 1983, **132**, 236.
[59] E. Palecek, P. Boublikova and F. Jelen, *Anal. Chim. Acta,* 1986, **187**, 99.
[60] G. Dryhurst, *Anal. Chim. Acta,* 1971, **57**, 137.
[61] G. Dryhurst, *Talanta,* 1972, **19**, 769.
[62] T. Yao, T. Wasa and S. Musha, *Bull. Chem. Soc. Jap.*, 1977, **50**, 2917.
[63] T. Yao, Y. Taniguchi, T. Wasa and S. Musha, *Bull. Chem. Soc. Jap.*, 1978, **51**, 2937.
[64] M. Moutet, R. Vallot and R. Buvet, *Bioelectrochem. Bioenerg.*, 1987, **18**, 137.
[65] R. A. Kenley, S. E. Jackson, J. C. Martin and G. C. Visor, *J. Pharm. Sci.*, 1985, **74**, 1082.
[66] G. C. Visor, S. E. Jackson, R. A. Kenley and G. C. Lee, *J. Pharm. Sci.*, 1985, **74**, 1078.
[67] J. M. Sequaris, P. Valenta and H. W. Nurnberg, *J. Electroanal. Chem.*, 1981, **122**, 263.
[68] V. Brabec, *Bioelectrochem. Bioenerg.*, 1981, **8**, 437.
[69] V. Brabec and G. Dryhurst, *J. Electroanal. Chem.*, 1978, **91**, 219.
[70] L. G. Karber and G. Dryhurst, *Anal. Chim. Acta,* 1979, **108**, 193.
[71] G. Dryhurst and L. G. Karber, *Anal. Chim. Acta,* 1978, **100**, 289.
[72] V. Brabec and G. Dryhurst, *Stud. Biophys.*, 1978, **67**, 23.
[73] F. Crespi, T. Sharp, N. Maidment and C. Marsden, *Neuroscience Lett.*, 1983, **43**, 203.
[74] R. D. O'Neill, M. Fillenz, R. N. Gruenwald, M. R. Bloomfield, W. J. Albery, C. M. Jamieson, J. H. Williams, and J. A. Gray, *Neuroscience Lett.*, 1984, **45**, 39.
[75] E. Szurley and A. Brajter-Toth, *Anal. Chim. Acta,* 1983, **154**, 323.
[76] J. Munson and H. Abdine, *Talanta,* 1978, **25**, 221.
[77] B. H. Hansen and G. Dryhurst, *J. Electroanal. Chem.*, 1971, **32**, 405.

[78] J. Wang, Internatl. *Clin. Prds. Rev.*, 1986, **5**, 50.
[79] W. F. Smyth, in *Electrochemistry, Sensors and Analysis*, M. R. Smyth and J. G. Vos (Eds), p. 29, Elsevier, Amsterdam, 1986.
[80] J. Wang and B. Freiha, *Bioelectrochem. Bioenerg.*, 1984, **12**, 225.
[81] H. Shiraishi and R. Takahashi, Bunseki Kagaku, 1985, **34**, 1.
[82] T. Iwamoto, M. Yoshiura and K. Iriyama, *J. Chromatogr.*, 1983, **278**, 156.
[83] K. Iriyama, M. Yoshiura, T. Iwamoto and Y. Ozaki, *Anal. Biochem.*, 1984, **141**, 238.
[84] K. Iriyama, M. Yoshiura, T. Iwamoto, T. Hosoya, H. Kono and T. Miyahara, *J. Liq. Chromatogr.*, 1983, **6**, 2739.
[85] K. Iriyama, M. Yoshiura and T. Iwamoto, *J. Liq. Chromatogr.*, 1985, **8**, 333.
[86] T. Iwamoto, M. Yoshiura, K. Iriyama, N. Towizawa, S. Kurihara and T. Aoki, *J. Liq. Chromatogr.*, 1986, **9**, 1503.
[87] R. J. Henderson and C. A. Griffin, *J. Chromatogr.*, 1984, **298**, 231.
[88] M. E. Dwyer and P. R. Brown, *J. Chromatogr.*, 1985, **345**, 125.
[89] A. M. Ghe, G. Chiavari and J. Evgenidis, *Talanta*, 1986, **33**, 379.
[90] D. Perret, *Biochem. Soc. Trans.*, 1985, **13**, 1067.
[91] T. Yamamoto, H. Shimizu, T. Kato, T. Nagatsu, *Anal. Biochem.*, 1984, **142**, 395.
[92] J. B. Kafil, H.-Y. Cheng and T. A. Last, 1986, **58**, 285.
[93] M. R. Smyth, *Analyst*, 1980, **105**, 612.
[94] E. C. Lewis and D. C. Johnson, *Clin. Chem.*, 1978, **24**, 1711.
[95] M. K. Halbert and R. P. Baldwin, *Anal. Chim. Acta*, 1986, **187**, 89.
[96] K. Mckenna and A. Brajter-Toth, *Anal. Chem.*, 1987, **59**, 954.
[97] E. Watanabe, S. Tokimatsu, K. Toyama, I. Karube, H. Matsuoka and S. Suzuki, *Anal. Chim. Acta*, 1984, **164**, 139.
[98] E. Watanabe, K. Ando, I. Karube, H. Matsuoka and S. Suzuki, *J. Food. Sci.*, 1983, **48**, 496.
[99] E. Watanabe, K. Toyama, I. Karube, H. Matsuoka and S. Suzuki, *Eur. J. Appl. Microb. Technol.*, 1984, **19**, 18.
[100] E. Watanabe, T. Ogura, K. Toyama, I. Karube, H. Matsuoka and S. Suzuki, *Enzyme Micro. Technol.*, 1984, **6**, 207.
[101] E. Watanabe, H. Endo, T. Hayashi and K. Toyama, *Biosensors*, 1986, **2**, 235.

3

Amino acids, peptides and proteins

3.1 INTRODUCTION

The analysis of amino acids and their metabolites is out of major importance in the biological and biomedical fields [1–10]. Reliable analytical methods are essential for determinations in complex matrices such as blood and urine; such methods need to possess both high sensitivity and selectivity. The monitoring of amino acids in protein hydrolysates is another area of considerable interest. In addition, procedures for the determination of peptides in biological tissues have gained increasing importance. The measurement of intact proteins is also of great interest since this can aid in a better understanding of fundamental processes involving protein/protein interactions; in addition, proteins may be used as markers for disease states.

Modern electrochemical techniques have great potential for the determination of amino acids and related compounds. With regards to electrochemical behaviour, inspection of the structures of the neutral α-amino acids (I–VI; where α refers to the fact that both the amino and carboxylic acid groups are on the same carbon atom) reveals that these do not contain any electroactive groups. However, a variety of sensitive and selective methods have been developed, some of which are based on derivitization procedures, and others of which involve the use of novel electrode materials. On the other hand, examination of the tyrosine (VII) and tryptophan (VIII) molecules shows that these contain electro-oxidizable phenolic and indole moieties respectively. Therefore, a number of direct voltammetric/LCEC methods have been developed for these compounds. A few other α-amino acids contain electroactive groups; for example, cysteine (IX) contains the electro-oxidizable thiol moiety, whereas cystine (X) contains the electro-reducible disulphide bridge.

The aim of this chapter is to describe representative applications where electroanalytical techniques have been used to solve analytical problems, including the types mentioned earlier. The division of the chapter into sections and subsections has not been particularly easy. For example, it was not usually practical to discuss each amino acid in isolation, because methods have generally been designed to monitor mixtures of these species; therefore, this would have led to repetition. Instead, the sections have been constructed according to the electrochemical techniques employed for the determinations; the discussion often considers a variety of amino acids, or amino acids and related compounds together (e.g. phenylalanine(XI) and L-dopa (XIII)).

3.2 ELECTROANALYSIS OF AMINO ACIDS, PEPTIDES AND PROTEINS

3.2.1 Methods involving voltammetry

3.2.1.1 Direct polarographic methods

Sulphur-containing amino acids: cystine and cysteine
The electrochemical behavior of cystine (X, RSSR), at a mercury electrode, was described by Lee [11]; it was reported that RSSR was reduced at the DME to produce cysteine (IX, RSH) as shown in (3.1).

$$RSSR+2e^-+2H^+ \rightleftharpoons 2RSH \tag{3.1}$$

In this study, fast linear-sweep polarography† was used and the supporting electrolyte consisted of 0.1 M $HClO_4$; the resulting polarographic peaks appeared at a potential of −0.43 V when the concentration of RSSR was 0.1 mM. It was also shown that if after the first scan the elctrode was poised for a short time at 0 V, prior to a second scan, a more anodic peak also appeared at $E_p=-0.08$ V. Apparently, this was due to the reduction of mercurous cysteinate (RSHg) as shown in (3.2)

$$RSHg+e^-+H^+ \rightleftharpoons RSH+Hg \tag{3.2}$$

The mercurous cysteinate was formed by oxidation of RSH (at 0 V) which was formed on the first cathodic scan; the oxidation process was:

$$RSH+Hg \rightleftharpoons RSHg+e^-+H^+ \tag{3.3}$$

In order to avoid the production of mercury cysteinate, the cell was left at open-circuit until just before scanning so that (3.3) could not occur.

Mairesse-Ducarmois *et al.* [12,13] have investigated the electrochemical characteristics of cysteine and cystine using d.c., a.c. and differential pulse polarography. For the simultaneous determination of these two compounds the authors selected DPP and a supporting electrolyte comprising 0.05-M $Na_2B_4O_7$–0.5-M KNO_3, pH 9.2. In this medium cysteine exhibited two anodic peaks with peak potentials at −0.535 V and −0.625 V versus SCE; cystine showed a peak with $E_P=-0.91$ V versus SCE. Under these conditions DPP permitted the determination of cysteine and cystine in the same solution down to 0.12 and 0.24 $\mu g\ cm^{-3}$ respectively.

† In this technique the complete voltage scan is performed on a single drop of mercury.

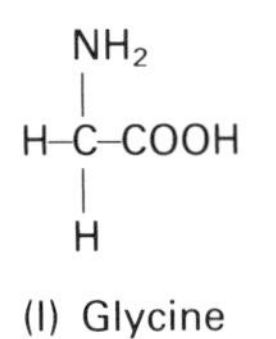

(I) Glycine

$$\begin{array}{c} NH_2 \\ | \\ CH_3{-}C{-}COOH \\ | \\ H \end{array}$$

(II) Alanine

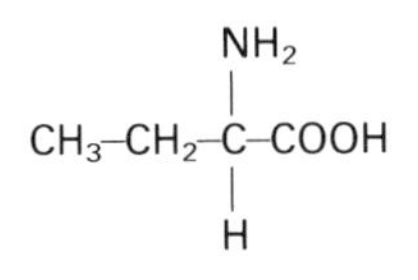

(III) Valine

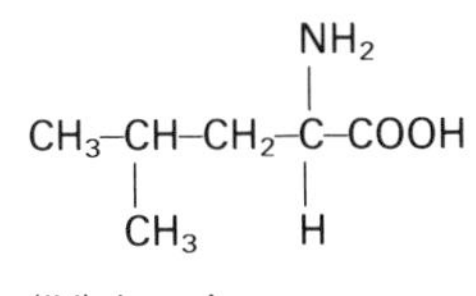

(IV) Leucine

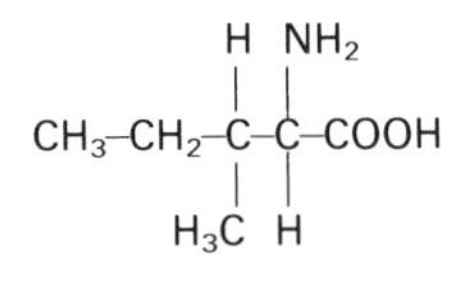

(V) Isoleucine

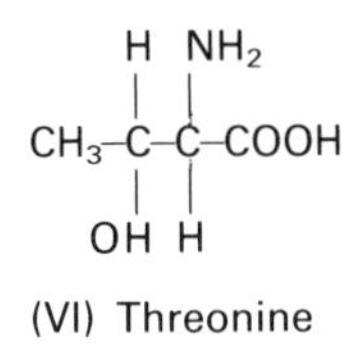

(VI) Threonine

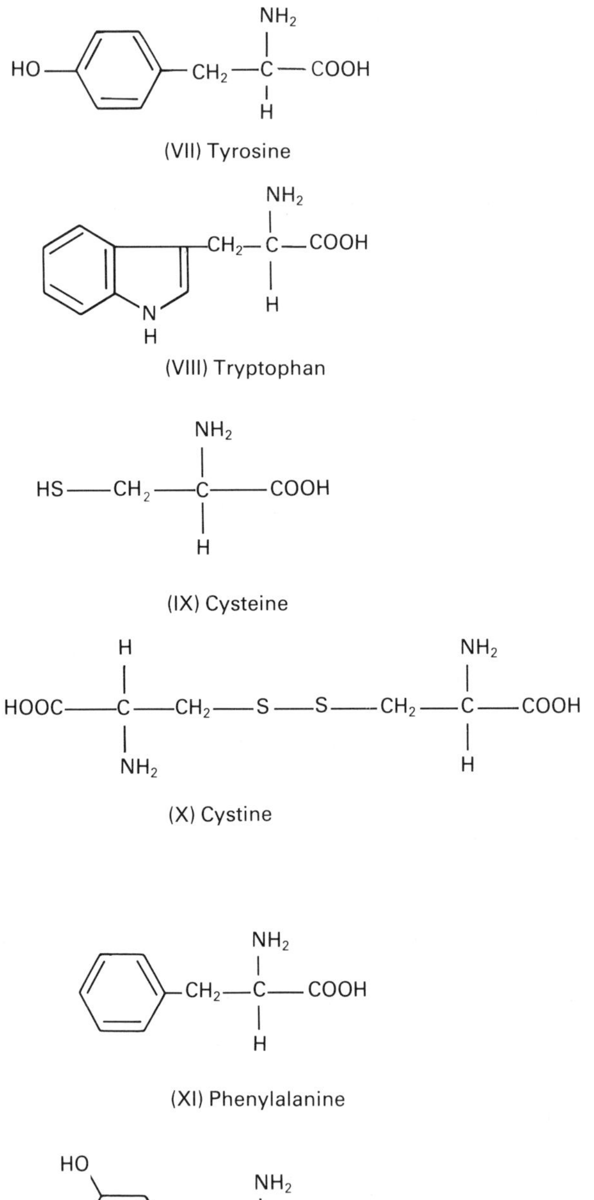

(VII) Tyrosine

(VIII) Tryptophan

(IX) Cysteine

(X) Cystine

(XI) Phenylalanine

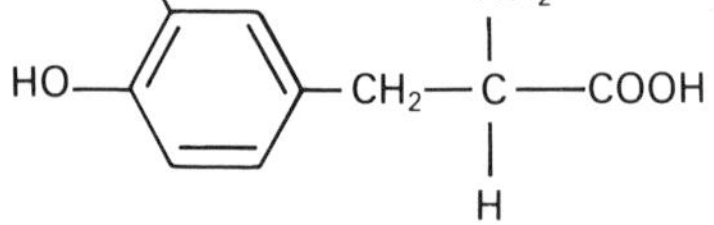

(XII) L-3,4-Dihydroxyphenylalanine (L-Dopa)

Interestingly, the DPP behaviour of cysteine has been exploited for the indirect determination of the sulphoxide metabolite 5-carboxymethyl-L-cysteine, which is not polarographically active [14]. The metabolite was hydrolysed in acid medium and the analysis was performed in Britton–Robinson buffer pH 4.0. A peak was observed at $E_P = -0.18$ V versus SCE which resulted from reaction (3.2).

Glutathione

Glutathione is a tripeptide with the chemical name (*N*-(*N*-L-γ-glutamyl-L-cysteinyl)-glycine; it can exist in nature in the reduced form (XIII, GSH) and the oxidized form (XIV, GSSG).

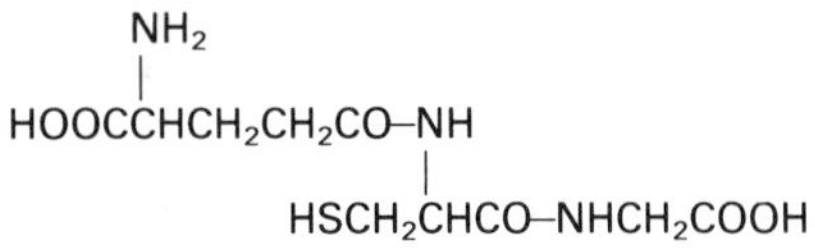

```
       NH2
       |
HOOCCHCH2CH2CO–NH
               |
          HSCH2CHCO–NHCH2COOH
```

(XIII) Reduced Glutathione (GSH)

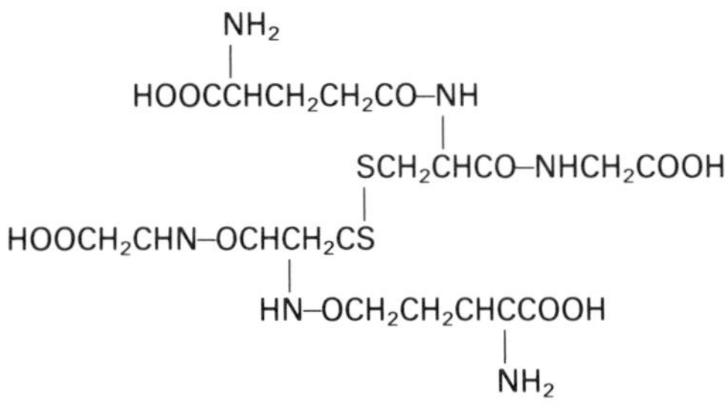

```
              NH2
              |
       HOOCCHCH2CH2CO–NH
                      |
                 SCH2CHCO–NHCH2COOH
                 |
HOOCH2CHN–OCHCH2CS
            |
          HN–OCH2CH2CHCCOOH
                       |
                       NH2
```

(XIV) Oxidized Glutathione (GSSG)

The d.c. polarographic behaviour of GSH and GSSG was described some time ago by Stricks and Kolthoff [15]. These workers reported that GSH produced two anodic waves at the DME. It was suggested that the normal wave corresponded to the formation of a mercurous compound in the pH region 1 to 10.5 as shown in (3.4):

$$GSH + Hg \rightleftharpoons GSHg + e^- + H^+ \tag{3.4}$$

The second wave was quite well-defined at an ionic strength of 1.0 M and this corresponded to the formation of GSHg; however, it was found that the characteristics of this wave were dependent on ionic strength [15]. In the same study, it was reported that the $E_{1/2}$ value for the normal wave was 55 mV more cathodic than the second wave at all pH values and GSH concentration investigated; this behaviour indicates that the latter is a prewave due to adsorption phenomenon (see below).

It was also reported [15] that GSSG exhibited only one cathodic wave at all pH values and concentrations studied.

The mechanism of reduction was described as:

$$GSSG + 2e^- + 2H^+ \rightarrow 2GSH \tag{3.5}$$

The effect of GSH concentration on the diffusion current was investigated over the range 5×10^{-4} M to 2×10^{-3} M; a linear relationship was found to exist over this range.

A more recent study on the electrochemical characteristics of reduced glutathione has been carried out by Patriarche and coworkers [16–18]; these workers employed d.c., a.c. and differential-pulse polarography for the investigation. In this study, it was shown that the product adsorbed onto the mercury surface following the oxidation of GSH. The DPP technique allowed determinations of down to 0.3 μg cm^{-3} but no applications were reported.

A simple, rapid polarographic method for the determination of GSH in whole blood was briefly described by Hart [19]. In this preliminary study, the author took 0.1 cm^3 of the sample and added it to 5.0 cm^{-3} of acetate buffer pH 4.3; to this was added 0.01 cm^{-3} of *n*-ocatanol, which served as an anti-foaming agent during deaeration. The resulting solutions were analysed by DPP and well-defined anodic peaks were obtained with an E_P of -0.25 V versus SCE (Fig. 3.1). A linear response was obtained over the concentration range 3.6 to 18 μg cm^{-3}. Although the author did not attempt a complete evaluation of the proposed method, some preliminary results on circulating GSH concentrations agreed with published values; this indicated that the method may be suitable for clinical investigations. Further studies are needed to ascertain whether all of the current measured at -0.215 V is due to reduced glutathione.

Proteins

Since proteins are built-up from α-amino acids (almost always in the L-configuration) it is perhaps not surprising that these large molecules possess similar electrochemical characteristics to the monomers.

Proteins which contain the disulphide group have been reported to exhibit polarographic activity due to reduction of the disulphide group [20,21]; the proteins investigated included insulin, trypsin ribonuclease and bovine serum albumin and these were reported to exhibit a single cathodic wave under acidic conditions. However, in some cases, in neutral and alkaline solutions a second wave appeared; this only occurred after the initial wave had reached a maximum value and indicates the presence of strong adsorption. Further investigations on insulin [22] have also led to the conclusion that this molecule undergoes strong adsorption onto the mercury electrode. Reduction of the molecule in the adsorbed state was stated to involve a four-electron process to form four sulphydryl groups. Similar behaviour has been reported by Stankovich and Bard [23] for the protein bovine serum albumin. A fuller description of the electrochemical behaviour of proteins has been given by Palecek [24].

Hertl [25] has carried out investigations on human serum and bovine serum albumin using DPP. In this study, human serum protein was found to give a DPP peak at -0.6 V versus Ag/AgCl in a supporting electrolyte comprising 0.05-M phosphate/0.1-M KCl. pH 7.0. In addition, bovine serum albumin (BSA) also exhibited similar DPP behaviour under the same conditions. It was shown that both

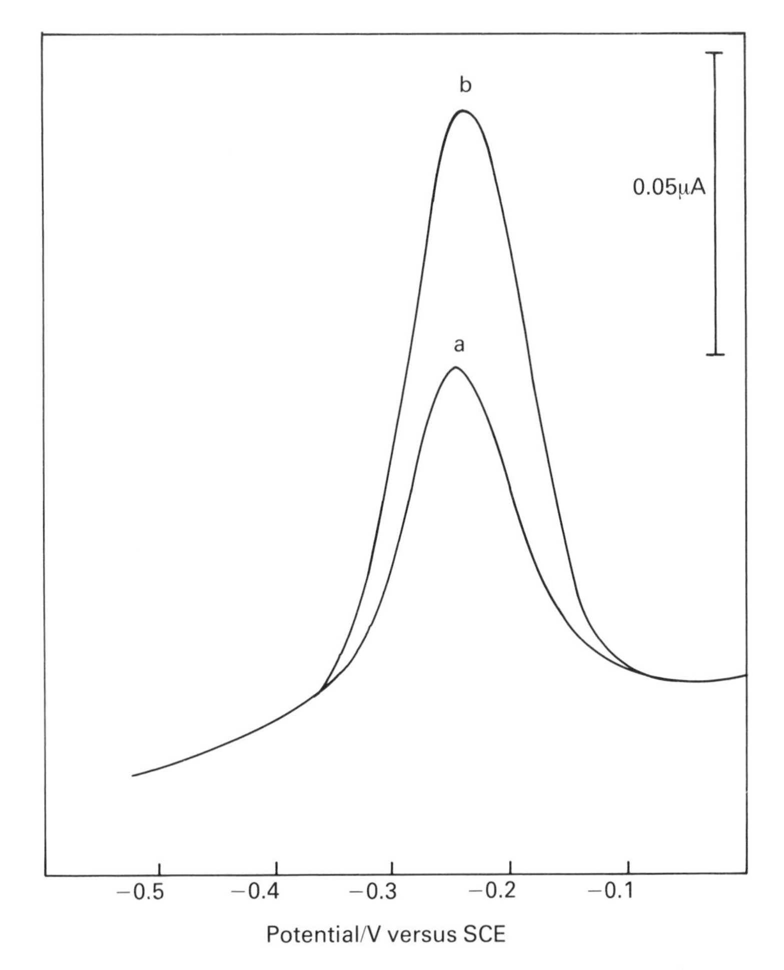

Fig. 3.1 — Differential pulse polarograms of: (a) 0.1 cm^3 of whole blood+5.0 cm^3 acetate buffer pH 4.3+0.01 cm^3 *n*-octanol; (b) a+0.04 cm^3 standard GSH (1040 $\mu g\ cm^{-3}$). Endogenous GSH concentration was 570 $\mu g\ cm^{-3}$.

BSA and normal serum produced linear calibration graphs by DPP; the slopes of these two graphs were essentially the same. Therefore, the polarographic assay for total human serum protein involved the use of BSA to generate a calibration graph. When this was used to calculate the protein in 10 human subjects the values agreed well with those obtained by a separate method (r=0.894); the range of the results was 3.8–8.8%. The method could also be applied to whole blood since cellular constituents did not interfere.

3.2.1.2 Polarographic methods using a dropping copper amalgam electrode

Various amino acids

An interesting approach to the polarographic determination of phenyalanine, tyrosine, methionine, glutamic acid, histidine and glycine has been reported by Perez and coworkers [26,27]. These researchers employed a dropping copper amalgam electrode† and the compounds mentioned gave anodic waves when the electrode was oxidized in their presence. For the analytical measurements the d.c. polarographic technique was used and the supporting electrolyte consisted of 0.5-M sodium chlorate; the working electrode was 2.2×10^{-3}-M Cu(Hg). Calibration graphs were linear in the range 1×10^{-3} to 1×10^{-4} M using the waves with $F_{1/2}=-0.28$ V; however for histidine the behaviour was slightly different. For this amino acid one anodic wave appeared at the isoelectric point with $E_{1/2}$ value of -0.2 V; as pH was increased two anodic polarographic waves appeared with the more negative at $E_{1/2}$ value at -0.38 V. The complexes formed during the oxidation process were considered to be those shown in Fig. 3.2; species (I) was formed when the amino

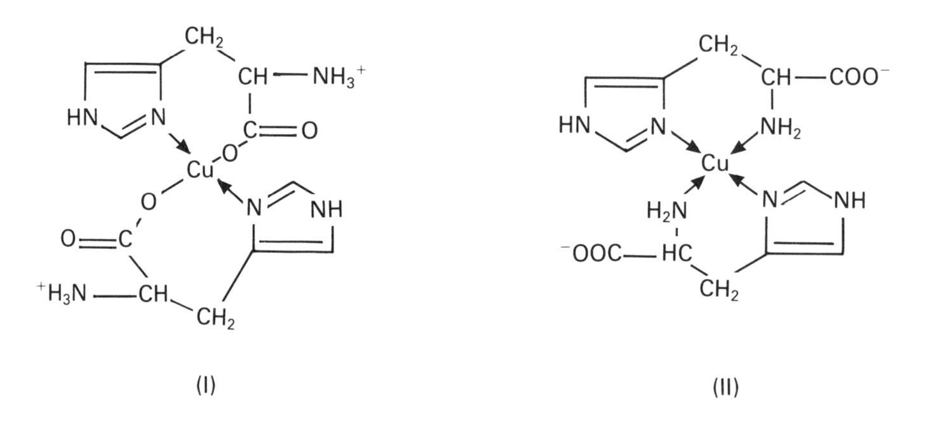

Fig. 3.2 — Complexes of histidine formed during the oxidation at a copper amalgam electrode. (Reproduced from [26] by permission of the copyright holders, Elsevier Science Publishers.)

group was protonated (amino acid was a zwitterion) and (II) when the nitrogen was unprotonated (amino acid was an anion). For quantitative measurements the anodic wave at $E_{1/2}=-0.38$ V was suitable since it increased proportionally as the histidine anion concentration increased. It was found that hydroxide ions interfered at low amino acid concentrations; the authors recommended separation of mixtures of amino acids prior to polarography because of the similarity in half-wave potentials.

3.2.1.3 Polarographic methods involving catalytic processes

Proteins

Catalytic polarographic currents are produced for proteins containing cysteine or cystine in solutions containing cobalt; this observation was made some time ago by Brdicka [28,29] and his name is now synonymous with polarographic methods using the phenomenon. A typical Brdicka response was obtained by Palecek [30] for serum

† This consists of metallic copper dissolved into mercury and is operated in the same manner as a DME.

albumin (70 μg cm^{-3}) in a medium containing 10^{-3} M $[Co(NH_3)_6]^{3+}$ with 0.1 M ammonium chloride and 0.1 M ammonium hydroxide (Fig. 3.3).

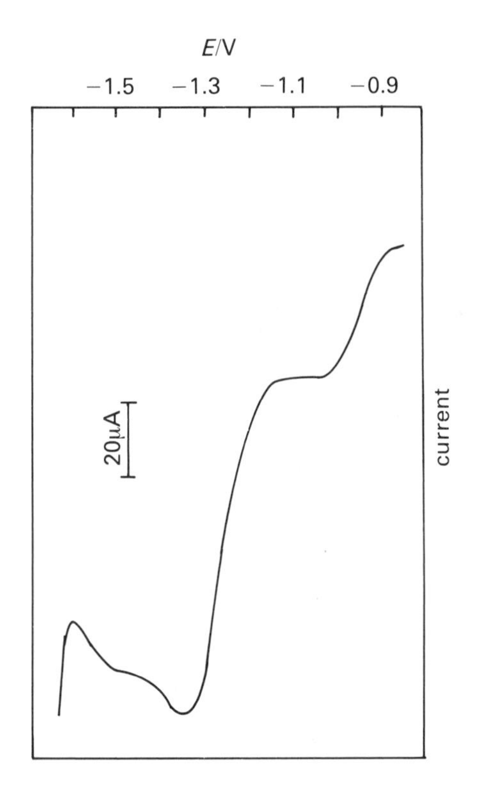

Fig. 3.3 — D.c. polarogram of albumin in 1×10^{-3}-M $[Co(NH_3)_6]^{3+}$ with 0.1-M NH_4Cl and 0.1-M NH_4OH. Albumin concentration 70 μg cm^{-3}; scan rate 2 mV s^{-1}; reference electrode was a mercury pool. (Adapted from [30].)

The mechanism responsible for the polarographic currents of proteins has not been fully explained, but a complex of protein with cobalt, from which hydrogen is catalytically evolved, plays a major part in the overall process. Following electrodeposition of metallic cobalt from the complex, the ionized sulphydryl groups take up protons from the acid component of the supporting electrolyte. These protons are reduced at a decreased hydrogen overvoltage, producing a catalytic current. The results of the studies of the mechanism from a number of research groups were summarized by Palecek [24].

In the remainder of this subsection some examples have been selected to illustrate the ways in which the technique may be exploited.

Palecek and Pechan [31] have carried out investigations for the determination of nanogram quantities of proteins using the DPP technique. Initial studies were carried out to ascertain the best conditions for the production of a well-developed DPP peak. Fig. 3.4 shows the DPP behaviour of serum albumin in several different

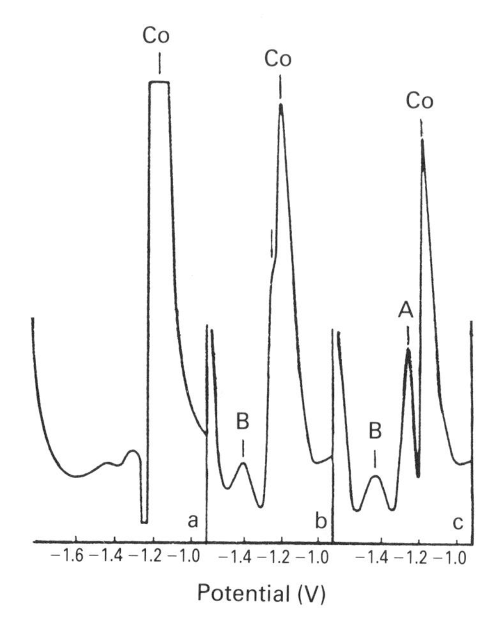

Fig. 3.4 — DPP of albumin in different media (bovine serum albumin (BSA) at a concentration of 1.5 $\mu g\ cm^{-3}$): (a) in 1×10^{-3}-M $[Co(NH_3)_6]Cl_3$ in 0.1-M ammonium chloride, ammonium hydroxide (Brdicka's cobaltic solution), containing 2×10^{-5}% Triton X-100; (b) 5.5×10^{-4}-M $[Co(NH_3)_6]Cl_3$ in 1.75-M ammonium chloride, 1.75-M ammonium hydroxide; (c) 6×10^{-4}-M $[Co(NH_3)_6]Cl_3$ in 1.0-M ammonium chloride, 1.0-M ammonium hydroxide with 2×10^{-5}% Triton X-100. The potentials were measured against a mercury pool at the bottom of the polarographic cell. (Reproduced from [31] by permission of the copyright holders Academic Press Inc.).

solutions; from this it is clear that a solution containing 6×10^{-4}-M $[Co(NH_3)_6]Cl_3$, 1-M NH_4OH,. 1-M NH_4OH with 2×10^{-5}% Triton X-100 gave suitable peaks for protein concentrations below 1 $\mu g\ cm^{-3}$. The two peaks A and B shown in Fig. 3.4 were shown to be linearly related to the concentration of serum protein in the range 0.05–1.0 $\mu g\ cm^{-3}$. The authors also investigated several other proteins under similar conditions, namely: insulin, ribonuclease, lysozyme and trypsin. All of these exhibited the two DPP peaks, but the magnitude was different for each protein mentioned. For analytical determinations in biological materials these workers recommended the use of wave B; in addition, they suggested the use of the standard addition technique to compensate for depression of the protein peak by other components of the matrix. In this investigation, the samples analysed included urine, saliva, wine, beer, vinegar, and lemon juice.

Many workers in Eastern Europe have used polarographic methods involving catalytic current for clinical investigations. In one such study serum and cerebro-spinal fluid were taken from normal subjects and from patients with mental disorders; it was stated that the resulting polarograms were significantly different [32]. Several reports have indicated that pronounced cathodic shift of the two polarographic Brdicka waves occurred for serum samples taken from subjects suffering with inflammatory and neoplastic disorders [33,34]. The same technique has also been used to investigate patients suffering with cirrhosis of the liver [35]. A number of other applications of the Brdicka reaction to clinical diagnosis have been reviewed by Chowdry [36].

Some interesting studies have been carried out on the polarographic behaviour of the virus tobacco mosaic virus (TMV) in Brdicka solution. The investigations carried out by Ruttkay-Nedecky's group [37–40] showed that at 0°C three polarographic waves were produced for the denatured TMV protein. It was stated that the whole virus particles behave in approximately the same way as the native protein [24].

The polarographic methods described above involved determinations in a stationary supporting electrolyte with normal, or micropolarographic cells; however, an automated continous-flow method for the determination of serum proteins has been reported by Alexander and Shah [41]. These authors described a flow-through cell for the application of a DME; by using DPP with a short drop time it was possible to run a rapid automated system for up to 120 samples/h. It was stated that with the Brdicka reagent, hexamine Co(III) chloride, a variety of serum proteins could be determined in the range 5–50 $\mu g\ cm^{-3}$.

Metal ions, other than those of cobalt, have been investigated for the polarographic determination of proteins; both the d.c. and DPP technique were used with a supporting electrolyte containing $NiSO_4$ with 0.3-M NH_4Cl and 0.3 NH_4OH for albumin studies [30]. In addition, Alexander *et al.* [42] have incorporated Rh(III) into the supporting electrolyte for polarographic studies on proteins.

Recently, a novel electrochemical immunoassay for bovine serum insulin that involves a combination of specific immuno-reaction with a catalytic polarographic method, has been described by Kano *et al.* [43]. This assay was applied to bovine serum test solutions containing intefering substances, such as bovine serum albumin and cysteine. Bovine insulin was separated from the test solution by adsorption on immunoadsorbents, which were prepared from anti-bovine insulin antiserum-immobilized gel; the analyte was then released by addition of urea. The DPP technique was used and the released insulin was added to the Brdicka solution to produce a catalytic current. The authors reported that the limit of detection for insulin was 40 $ng\ cm^{-3}$.

Methods involving Brdicka catalytic currents have been reviewed by several authors [44–47].

3.2.1.4 Polarographic methods following derivatization

Various amino acids

Al-Hajjaji [48] has investigated the d.c. polarographic reduction of 2,4,6-trinitrobenzene-l-sulphonic acid (TNBS) and some 2,4,6-trinitrophenyl-amino acid derivatives. Apparently, 2,4,6-TNBS reacts stoichiometrically and rapidly in alkaline conditions with most amino acids yielding 2,4,6-trinitrophenyl (TNP) derivatives [49]. Al-Hajjaji [48] carried out the trinitrophenylation reaction according to Snyder and Sobocinski [50]. This was performed in a total volume of 10 cm^3 in a 25 cm^3 volumetric flask using 2.5 cm^3 of aqueous 2,4,6-TNBS solution (0.01 M) followed by 0.25–1.0 cm^3 of amino acid solution (0.01 M). At zero time 5.0 cm^3 borate buffer

(0.025 M, pH 9.3) was added and the reaction mixture was diluted to 10 cm^3 with water; the reaction time was 30 min. The solution was then made up to volume with sodium sulphite (0.017 M) in phosphate buffer (0.1 M, pH 6.4); the final pH was found to be about 7. The number of electrons involved in the reduction of the derivatives was determined in phosphate buffer pH 7.0 (0.5 M); this was found to be 12, which could be explained if the three nitro groups in these species were reduced only to the hydroxylamine groups. For quantitation the presence of sulphite was necessary since it formed complexes with the TNP-amino acid derivatives; this produced well-defined waves even in the presence of excess reagent. The half-wave potentials for serine, threonine, glycine, and histidine were found to be -1.0 V, -1.007 V, -1.021 V and -0.949 V respectively. These waves were used to determine the amino acids investigated in the range 1 to 4×10^{-4} M; the precision was 2% (n=10).

Further investigations have been carried out by Al-Hajjaji [51] to determine a number of amino acids following a modification of the above procedure; furthermore, the more sensitive DPP technique was used for quantitation. In this study the DPP behaviour of 2.4.6-TNP derivatives was studied in the presence of sulphite; all of these produced a polarographic peak for their complexes with sulphite (1×10^{-2} M) in pH 8.0 phosphate buffer (0.05 M)–0, 1-M potassium chloride (Table 3.1). A

Table 3.1 — E_P and i_P values for TNP-amino acid sulphite complexes (1×10^{-5} M) in pH 8.0 (0.05 M)/ 0.5 M KCl/0.01 M Na_2SO_3. (Adapted from [51])

Compound	E_P/V	i_P/nA
Histidine	− 924	77
Glycine	−1000	50
Serine	− 984	48
Threonine	− 996	52
Glutamic acid	−1016	57
Asparagine	−1024	51
Lysine	−1016	46
Acetyllisine	− 952	68
Histamine	− 868	55
Arginine	− 870	12

5-min reaction time at room temperature (or 50°C for lysine) and pH 10.5 using 1×10^{-4}-M TNBS provided the optimal conditions for the determination of 5×10^{-6}-M to 2.5×10^{-5}-M amino acids; Fig. 3.5 shows the differential-pulse polarograms obtained for glycine over this concentation range. The relative standard deviation for the determination of 1×10^{-5}-M glycine was 1% (n=5).

The author [51] suggested that the method may be applied to studies of protein degradation, since proteins and peptides did not give polarographic peaks under the described conditions. Some selectivity for different amino acids could be obtained by solvent extraction, but generally peak resolution was not sufficient for complex mixtures of the amino acids investigated (Table 3.1).

Improved selectivity may of course be achieved by simply incorporating a

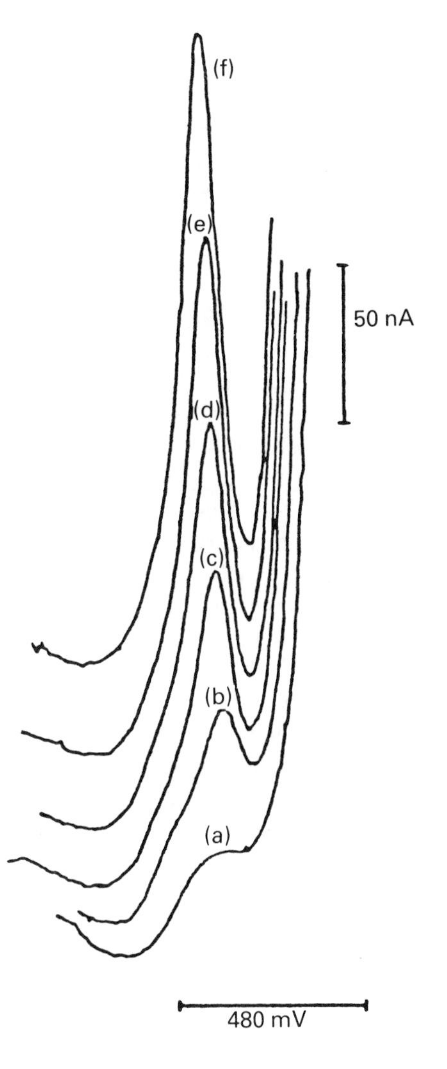

Fig. 3.5 — Polarograms of glycine used for obtaining calibration graphs: (a) 0.0 M; (b) 5×10^{-6} M; (c) 1×10^{-5} M; (d) 1.5×10^{-5} M; (e) 2×10^{-5} M; (f) 2.5×10^{-5} M. Conditions: 1×10^{-4}-M TNBS and 0.01-M sulphite at pH 8.0 (0.5-M KCl); Initial potential −0.8 V. (Adapted from [51].)

chromatographic step prior to polarography. This approach was adopted by Ramos *et al.* [52] for the determination of tyrosine and phenylalanine. These two compounds were separated from amino acid mixtures by thin-layer chromatography on cellulose sheets; butanol–acetic acid–water (4 : 1 : 1) was used as the mobile phase and the compounds were visualized under UV radiation. The zones corresponding to the two amino acids were removed and extracted with 20 cm^3 of nitric acid (0.3 M for tyrosine and 1.0 M for phenylalanine); these solutions were heated under reflux for 15 min (for tyrosine) 0r 20 min (for phenylalanine) and then excess acid was neutralized with NaOH. The resulting solutions were diluted to 100 cm^3 with water; DPP was then carried out on solutions containing ammonium chloride–ammonia (pH 9.5) as the supporting electrolyte. The calibration graphs were almost rectilinear for 01.–300 $\mu g\ ml^{-1}$ of tyrosine and rectilinear for phenylalanine; the respective detection limits were 0.05 and 0.08 $\mu g\ ml^{-1}$. The coefficient of variation of the method for the former amino acid was 2.5% (n=10).

3.2.1.5 Methods involving stripping voltammetry

Amino acids and some related compounds

As mentioned in earlier sections (Chapters 1 and 2), methods involving stripping voltammetry are particularly suited to low-level (below 1 $\mu g\ cm^{-3}$) analytical determinations. A number of compounds that contain sulphur have been found to be amenable to cathodic stripping voltammetry (CSV). The first step usually involves oxidation of the analyte to form a mercury-containing compound at modest positive or negative potentials; this is then preconcentrated at the stationary mercury electrode via adsorption. The second step involves stripping the adsorbed layer back into solution by the application of a cathodic scan; thus the mercury–sulphur complex undergoes a reduction reaction and the resulting signal is used to quantitate the original species. Several reviews on the principles and applications of the technique have been published [53–55].

Since cysteine (IX, RSH) contains a thiol moiety, it is perhaps not surprising that several independent investigators have reported methods involving cathodic stripping voltammetry for the determination of this important amino acid.

In one study, Stock and Larson [56] used a supporting electrolyte containing 1-M acetate buffer pH 4.6 together with an HMDE. Preconcentration was performed at a potential of +0.05 V versus SCE; the deposition period consisted of 1 min with stirring (50 rpm) followed by 1 min quiescence. Presumably, during this stage the reaction shown in Eq. (3.3) proceeded to produce RSHg, or possibly $(RS)_2Hg$ was formed [56]. The stripping step was performed using LSV with a scan rate of 20 mV s^{-1}. Under these conditions a cathodic peak was obtained with E_P=−0.175 V; this may be the result of a reaction similar to that shown in Eq. (3.2). The authors suggested that the cathodic peak could be used to determine RSH in the range 0.4–1.2 μM.

In another study, Berge and Jeroschewski [57] indicated that the linear calibration graphs, obtained by CSV, extended to 3 μM. These authors employed both phosphate and acetate buffers in these investigations. Similar results to these were also described by Vydra *et al.* [58].

In a recent study, a slightly different approach was investigated for CSV of the amino acids cysteine and tryptophan (VIII). The workers employed a copper-based mercury film electrode which was compared to an HMDE and a copper amalgam electrode [59]. It was reported that lower detection limits were obtained with the copper-containing electrodes than with the HMDE. It was also stated that the copper-based mercury film electrode was easier to prepare and maintain than the copper amalgam electrode; with the former, the detection limits for cysteine and tryptophan were 10 nM.

Copper ions have also been found to give enhanced sensitivity in the CSV determination of penicillamine (XV) [60]. In this investigation, the penicillamine was used to measure indirectly four antibiotic penicillin compounds (XVI–XIX). These latter compounds were first hydrolysed to penicilloic acid (XX) (Fig. 3.6); after the hydrolysis step accumulation was performed at an HMDE at −0.1 V versus SCE in a pH-4.6 buffer containing an excess of Cu^{2+}.

During the preconcentration step, the penicilloic acid was further degraded and an adsorbed layer of a copper complex of penicillamine was formed. This complex was subsequently reduced at about −0.4 V in the stripping step to produce a copper amalgam and free thiol:

$$CuRS_{(ads)} + H^+ + e^- \rightleftharpoons CuHg + RSH_{(sol)} \qquad (3.6)$$

Concentrations down to about 10^{-8} M could be measured by linear-sweep voltammetry; however, down to 2×10^{-10} M could be measured using differential-pulse CSV and a deposition time of 10 min.

Proteins

Wang *et al.* [61] have recently reported that the disulphide proteins trypsin and chymotrypsin could be determined in trace quantities using adsorptive stripping voltammetry (AdSV). In this study, controlled interfacial accumulation was performed onto an HMDE and DPV was used in the stripping step. The adsorbed proteins yielded well-defined cathodic peaks at about −0.3 V versus Ag/AgCl; the authors indicated that this was due to reduction of the disulphide moieties in the molecules. The adsorptive stripping response was evaluated with respect to preconcentration time and potential, pH, concentration, stripping mode and other variables. It was shown that short preconcentration periods permitted determinations in the range sub-micromolar to nanomolar. Electroactive surfactants were found to exert only minor effects. The reproducibility of the determination (at the 3×10^{-7}-M level) may be as low as 0.8% (RSD); it was suggested that this method possesses potential in both research and clinical laboratories.

Smyth and coworkers [62] have recently investigated the possibility of monitoring BSA using AdSV at an HMDE. The supporting electrolyte was 0.05 M phosphate buffer pH 7.4 and an accumulation potential of +0.15 V (versus Ag/AgCl) was found to be appropriate. When an accumulation time of 120 s was used the calibration graph, based on the peak at −0.55 V, was linear over the range $0.2–1.5\times10^{-8}$ M. The same research group have also studied the interaction of Conconavalin A (Con A) with mannose using this technique [63]. Con A exhibited two AdSV peaks at −0.21 V (peak A) and −0.53 V (peak B), which increased in magnitude with increasing accumulation time (Fig. 3.7).

Both peaks gave linear plots for current versus accumulation time up to 300 s; at longer times the surface of the HMDE became fully covered with adsorbing protein and the current reached a limiting value. It was suggested that peak A may be due to adsorption of protein at the electrode surface, whereas peak B could be due to a faradaic process involving the reduction of disulphide linkages, or protonated amino functions in the protein structure. Interestingly, when mannose was added to a 9.8×10^{-7}-M solution of Con A in the electrochemical cell it was shown that the peak

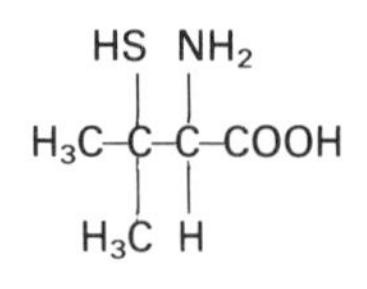

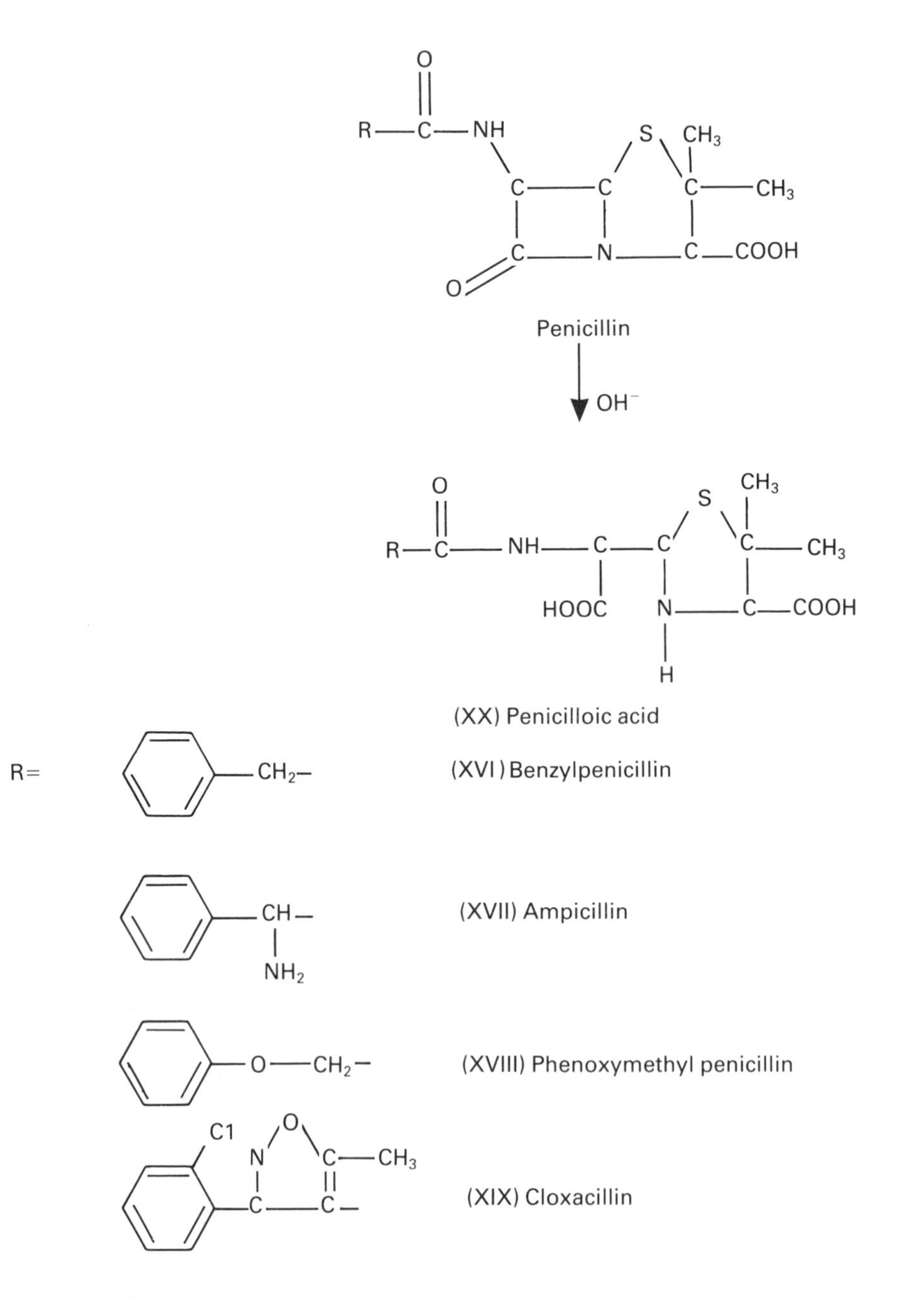

Fig. 3.6 — Hydrolysis of some penicillin compounds. (Adapted from [60].).

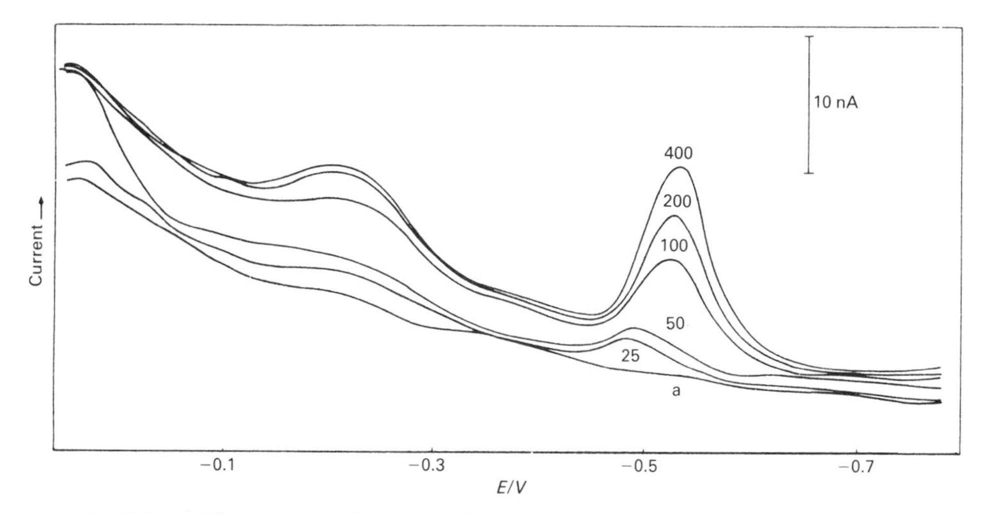

Fig. 3.7 — Effect of accumulation time (s) on the differential pulse AdSV behaviour of concavalin A. Accumulation potential, +0.05 V; scan rates mV s^{-1}; area of hanging mercury drop, 0.025 cm^2; pulse amplitude 50 mV, a=blank. Reproduced from [63] by permission of the copyright holders, Royal Society of Chemistry.).

B current increased linearly with respect to mannose concentration. The authors suggested that the results of their study could form the basis for the development of a biosensor for mannose.

Further studies by the same group have been concerned with the application of AdSV to monitor the interaction of human serum albumin (HSA) with anti-HSA [64], and mouse immunoglobulin G (IgG) with anti-mouse IgG [65]. In addition, these workers have also studied the interaction of human chorionic gonadotrophin (HCG) with each of its specific antibodies in solution using AdSV [66].

3.2.1.6 Methods involving voltammetry with unmodified carbon electrodes

Aromatic amino acids

Investigations have been carried out by Linders *et al.* [67] to develop a voltammetric procedure for measuring tryptophan in mixtures; apparently the anodic signals were produced by the oxidation of tryptophan to oxyindolalanine (XXI, Fig. 3.8).

These workers studied both vitreous carbon and carbon paste electrodes under a variety of solution conditions. For analytical purposes the optimum electrolyte was reported to be 0.1-M phosphoric acid; with this solution the detection limits were 1 μM and 0.25 μM at vitreous carbon and carbon paste respectively, and the respective calibration graphs were rectilinear for 3—200-μM and 0.5–30-μM tryptophan. This amino acid could be determined in the presence of serotonin (XXVIII) (5-hydroxytryptamine) owing to the differences in their oxidation potentials; Fig. 3.9 shows the differential-pulse voltammograms of the two compounds. Serotonin exhibits peaks at +0.59 V and +1.22 V, which are the result of oxidation at the 5-hydroxy position and the indole moiety respectively [68]. It was also possible to measure tryptophan in the presence of 5-hydroxytryptophan (XXX) but not in the presence of tyrosine, tryptamine or indole.

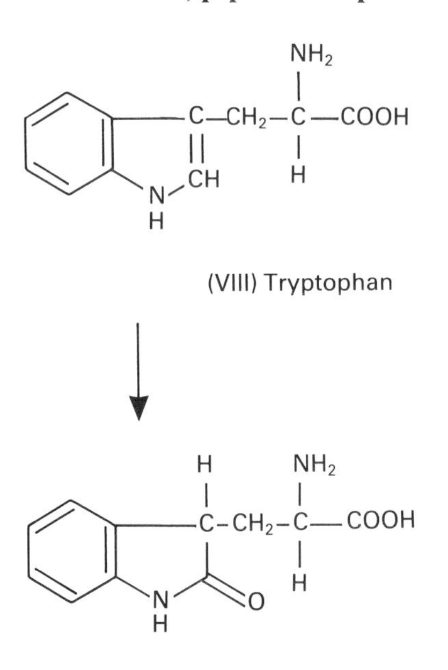

Fig. 3.8 — Oxidation of tryptophan at a carbon electrode. (Adapted from [67].).

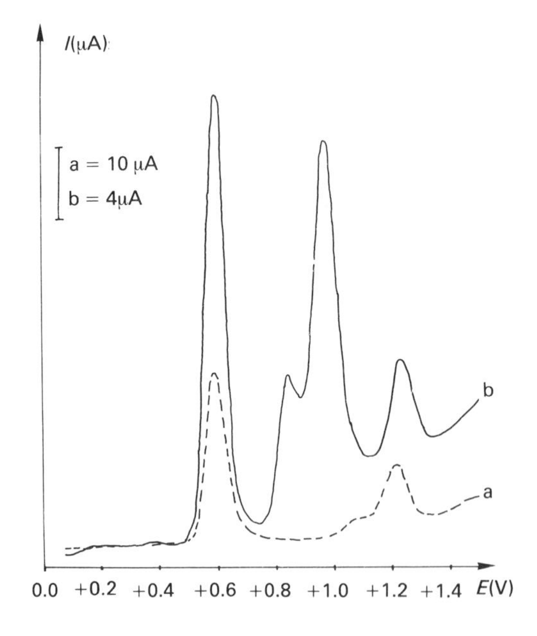

Fig. 3.9 — DPV of: (a) serotonin; (b) tryptophan and serotonin. Supporting electrolyte 0.1-M phosphoric acid; concentration of compounds 6×10^{-4} M; pulse amplitude 25 mV; scan rate 10 mV s^{-1}. (Reproduced from [67] by permission of the copyright holders.).

NH_2
|
$HS{-}CH_2{-}CH_2{-}CH{-}COOH$

(XXII) Homocysteine

$HNCOCH_3$
|
$HS{-}CH_2{-}CH{-}COOH$

(XXIII) *N*-acetylcysteine

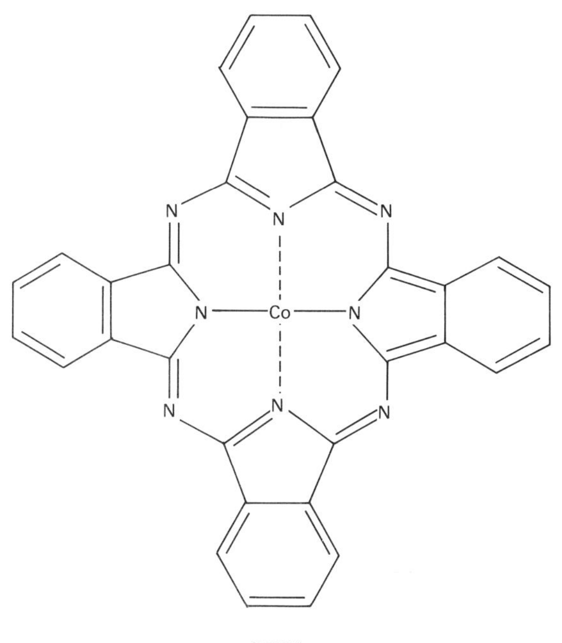

(XXIV)

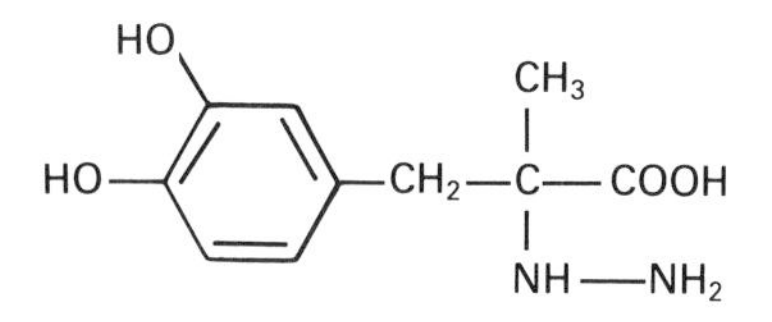

(XXV) Carbidopa

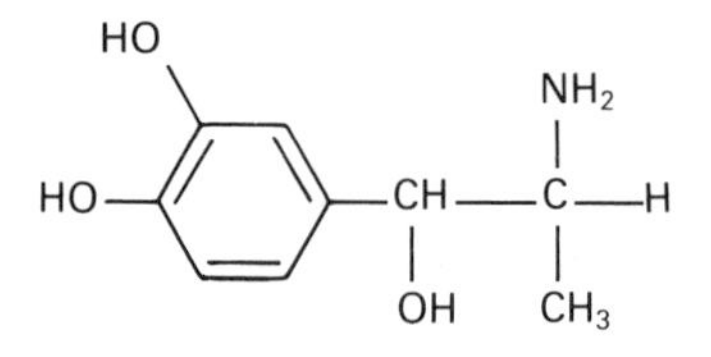

(XXVI) α-methylnoradrenaline

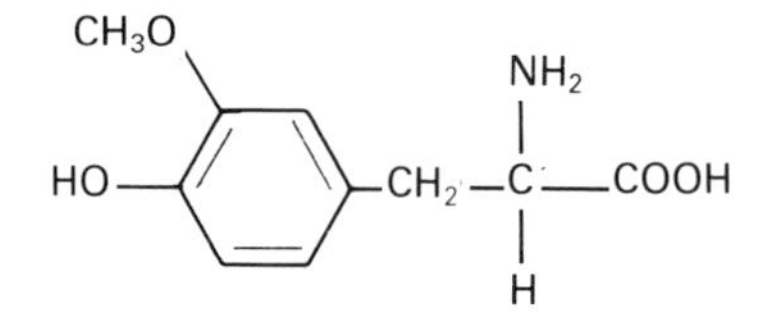

(XXVII) 3-*O*-methyl-dopa

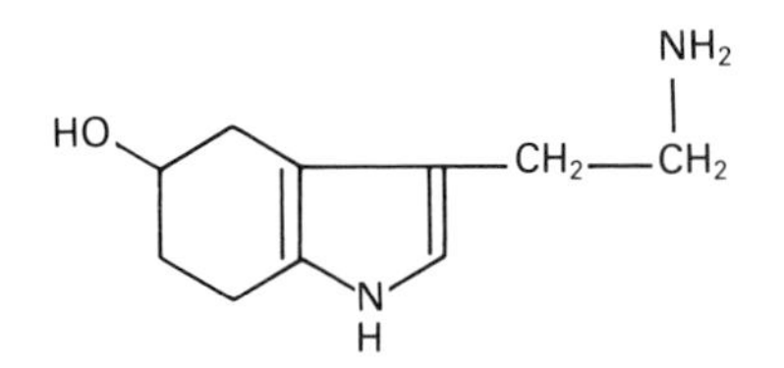

(XXVIII) 5-Hydroxytryptamine (Serotonin)

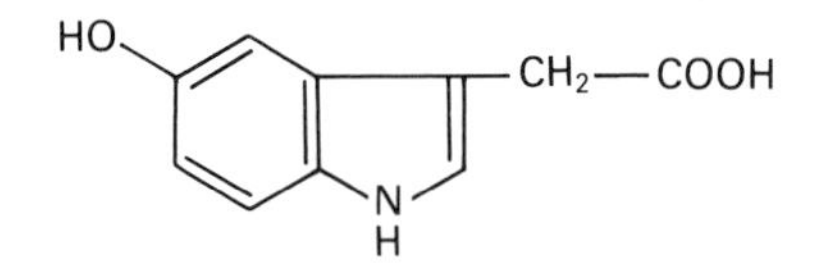

(XXIX) 5-Hydroxyindoleacetic acid

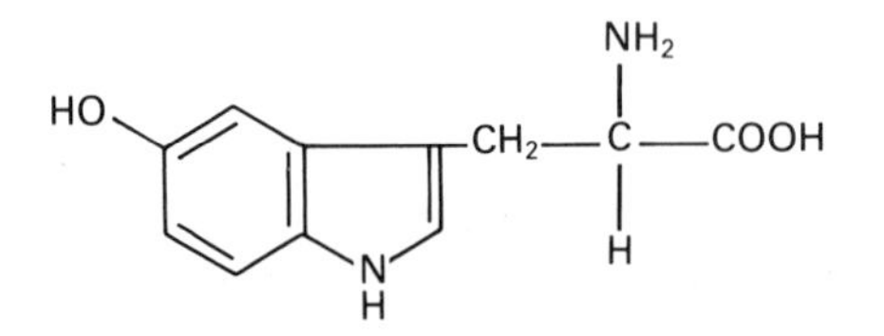

(XXX) n5-Hydroxytryptophan

In a recent study, tyrosine and L-dopa were studied by DPV using a paraffin wax-impregnated graphite electrode [69]; for this investigation the workers used a scan rate of 10 mV s^{-1} and a modulation amplitude of 100 mV. In a supporting electrolyte consisting of 10-mM sulphuric acid these two amino acids were reported to give separate anodic peaks; this allowed determinations in the ranges 0.05–50 and 1.0–90 μM for tyrosine and L-dopa respectively. It was also suggested that the substances could be converted into isoindoles by reaction with 3-mM phthalaldehyde and 1-mM mercaptosuccinic acid in borate buffer pH 9.2. The reaction products could be used to measure 10 to 100 μM of total amino acids. The selective determination of L-dopa and determination of the sum of tyrosine and L-dopa, after their conversion into isoindoles, has been applied to their determination in fermentation processes [69].

3.2.1.7 Methods involving voltammetry with chemically modified electrodes

Sulphur-containing amino acids and related compounds

In recent years, electrodes possessing specific chemical functionalities, intentionally linked to their surface, have been found to possess some advantages over conventional electrode substrates in electroanalysis. One important property of CMEs is their ability to catalyse the oxidation, or reduction, of analytes in solution that exhibit high overvoltages at unmodified surfaces; these species would not normally be suited to quantitation by conventional electrochemical approaches. In this subsection several studies are described to illustrate the potential that voltammetric methods have for the determination of some sulphur-containing amino acids and related compounds.

Cyclic voltammetric studies have been carried out by Halbert and Baldwin [70] on cysteine (IX), homocysteine (XXII), *N*-acetyl cysteine (XXIII) and glutathione (XIII), at both unmodified and modified carbon paste electrodes; the latter contained 2% cobalt phthalocyanine (CoPC, XXIV) mixed into the carbon paste material. It was shown that all four compounds exhibited peaks between +0.75 and +0.85 V at the CoPC-containing electrode, which were due to the activity of cobalt as an electron mediator, i.e. Co(III) is chemically reduced to Co(II) by GSH with a consequent increase in the Co(II) oxidation peak. However, none of these compounds gave a peak below +1.0 V at the unmodified carbon paste electrode. The modified electrode has been exploited by Halbert and Baldwin for the determination of cysteine and GSH using LCEC [71] (discussed later in section 3.2.2.2), but they do not appear to have studied the possibility of using a voltammetric approach.

The present author and coworkers have also been interested in the possibility of measuring GSH with CoPC-containing carbon paste electrodes. In one of our recent investigations [72] we carried out systematic studies on GSH using cyclic voltammetry at both unmodified and CoPC-containing carbon paste electrodes; the effect of pH, ionic strength of the buffer and percentage of the modifying agent on the response to GSH was investigated. Fig. 3.10 shows the cyclic voltammograms obtained for reduced glutathione (3.88×10^{-4} M) at both unmodified and modified electrodes in Britton–Robinson buffer pH 12.

Clearly, no anodic peaks were observed for GSH at the former electrode over the range studied; whereas a new peak (peak 1, Fig. 3.10a) appeared when the CoPC electrodes was changed from plain buffer solution (Fig. 3.10b) to one containing

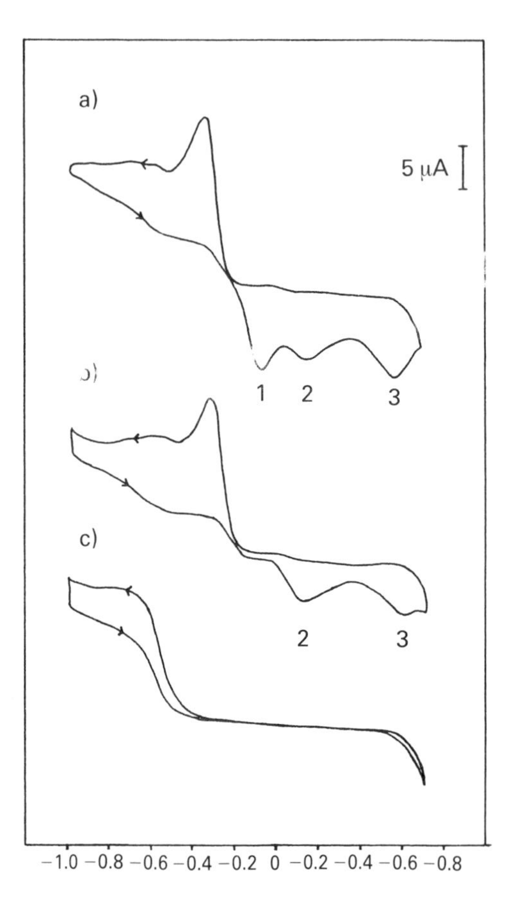

Fig. 3.10 — Cyclic voltammograms of: (a) 3.88×10^{-4}-M GSH at a modified carbon paste electrode containing 2% CoPC; (b) the modified electrode in plain buffer; (c) 3.88×10^{-4}-M GSH at an unmodified carbon paste electrode. In each case the supporting electrolyte was Britton–Robinson buffer pH 12.0. Scan rate, 50 mV s^{-1}. (Reproduced from [72] by permission of the copyright holders, Royal Society of Chemistry.)

GSH. As far as we know this behaviour has not been reported previously by other workers. One explanation for the appearance of peak 1 is that GSH may chemically reduce Co(II) to Co(I) and that peak 1 is the result of the re-oxidation of Co(I) to Co(II); this is shown schematically in Fig. 3.11.

The appearance of peak 2 is considered to be the result of the oxidation of Co(II) to Co(III), which was found to occur in the absence and in the presence of GSH; it should be noted that peak 2 increases with GSH concentration and this has also been reported by Halbert and Baldwin [70].

The author and coworkers have further investigated the possibility of carrying out quantitative analysis for GSH by DPV with the CoPC-containing carbon paste

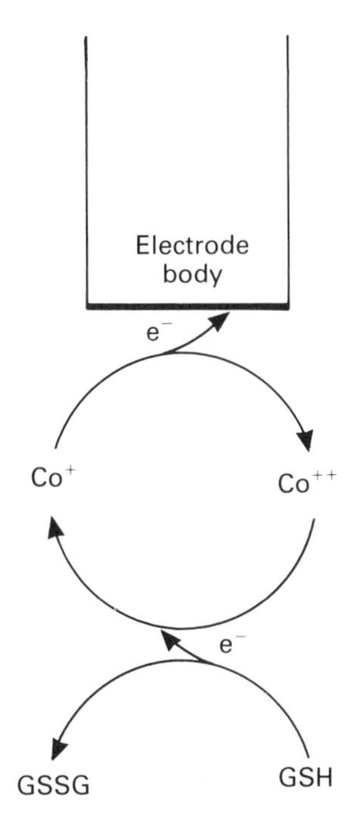

Fig. 3.11 — Electrode process for the formation of peak 1, showing the oxidation of GSH at the modified electrode. (Reproduced from [72] by permission of the copyright holders, Royal Society of Chemistry.)

electrode [72]. For this study a supporting electrolyte containing 0.1-M phosphate buffer pH 5.0 was found to be optimal; at higher pH values GSH was found to be unstable. Calibration graphs were constructed for GSH using peaks 1 and 2. For peak 1 (E_P=+0.27 C versus SCE) the graph was found to be linear over the range 2.5×10^{-6} to 6.25×10^{-5} M; for peak 2 (+0.75 V) linearity was obtained over the range 1.25×10^{-4} to 2.5×10^{-3} M GSH. These different calibration ranges suggest that peak 1 may be suitable for the determination of the low GSH levels present in human plasma; however, for the higher GSH concentrations present in whole blood the measurement of peak 2 may be more appropriate. It should be remembered that with these carbon paste electrodes only one measurement is made on each electrode; this is then removed and replaced with fresh material for the next measurement.

The author and colleagues also reported on the development of the CoPC-containing carbon electrode which is reusable [72]; the constituents of this material (epoxy resin/graphite mixture) are described later (Section 3.2.2.). The electrode was fabricated by inserting the material into the end of a 3-mm-diameter hollow glass tube; electrical contact was made by pushing a wire down the inside of the rod and into the back of the mixture. The epoxy resin/CoPC mixture was then cured to produce a hard polishable surface; after a suitable polishing procedure this was ready for use. This electrode was actually investigated as a possible amperometric sensor for GSH determination using stirred solutions and a conventional voltammetric cell.

In this mode of operation it was found that a potential of +0.5 V was suitable for the detection of GSH; a calibration graph was found to be linear over the range 3.9 μg cm^{-3} to 1.69 mg cm^{-3} and the limit of detection was 10 ng cm^{-3}. Several possible naturally occurring interferences were studied; ascorbic acid, cysteine and uric acid gave a response at the electrode but methionine did not. This would suggest that for applications in biological fluids some means of improving the selectivity would be required; this possibility is currently being investigated and will be reported at a later date. It should be added that the amperometric method described here while lacking in the necessary specificity for the analysis of human blood may still be applicable to the determination of GSH in less complex samples.

It should be noted that the reusable CoPC electrode may also be operated as an electrode for use with the DPV technique. In this case, it is merely necessary to wash the electrode in between runs with water and dry with a tissue.

It has been suggested by Halbert and Baldwin [70] that 2-mercaptoethanol, 3-mercaptopropionic acid, the antitumor drug 6-mercaptopurine, the antihypertensive agent Captopril and penicillamine all respond at CoPC-containing carbon paste electrodes. Therefore, the electrodes that we have developed, as well as those reported by Halbert and Balwin [70,71], could have wide application.

3.2.2 Methods involving liquid chromatography with electrochemical detection (LCEC)

3.2.2.1 LCEC methods using unmodified carbon electrodes

Tyrosine, tryptophan and their metabolites

Tyrosine (VII) and the catechol α-amino acid known as L-dopa (XII, 3,4-dihydroxyphenylalanine) undergo oxidation at the aromatic hydroxy groups [73]; a variety of LCEC methods have exploited this behaviour for trace analysis in various samples.

L-Dopa has been used in the treatment of Parkinson's disease and Causon *et al.* [74] have developed a method for measurement of therapeutic levels. The chromatographic separation was carried out with an Ultrasphere octyl column (5 μm); the mobile phase consisted of citrate–phosphate buffer pH 3.1, containing 2×10^{-3} M EDTA (disodium salt), 6.5×10^{-3}-M 1-octanesulphonic acid and 14% v/v methanol. This provided good separation of L-Dopa from carbidopa (XXV, L-α-hydrazino-3,4-dihydroxyl-L-methylbenzenepropanoic acid) present in the plasma of Sinemet-treated Parkinsonian patients, and also from α-methylnoradrenaline (XXVI), which was the internal standard; in addition there was no interference from the metabolite 3-*o*-methyldopa (XXVII, 3-OMD). Detection was carried out with a thin-layer amperometric cell containing a glassy carbon working electrode; this was set at +0.72 V versus Ag/AgCl. Prior to injection of the analyte onto the column, a clean-up step was performed by adsorption of the analyte onto Alumina; this was then desorbed using 0.1-M orthophosphoric acid. The plot of peak height ratio versus plasma L-dopa concentration was found to be linear over the range 50–2000 ng ml^{-1}. The absolute recoveries of the L-dopa and internal standard were found to be 72% and 70% respectively (n=6). An absolute detection limit of 25 ng ml^{-1} was calculated based on a signal-to-noise ratio of 2. Since the therapeutic range was reported to be 0.1–3.0 μg ml^{-1}, the sensitivity of the method was clearly satisfactory. The precision of the method was found to be 1.8% (intra-assay) and 4.3% (interassay), which was studied over a one-month period. The authors suggested that the proposed method would be suitable for the study of the effects of L-dopa on the cerebral blood flow and metabolism of patients with Parkinson's disease.

Several other methods have also been reported for the measurement of L-dopa levels following therapeutic treatment with this substance [75,76].

LCEC methods have been reported for the determination of L-dopa that also have the advantage of simultaneously monitoring the important metabolite 3-OMD.

Beers *et al.* [77] described an LCEC method which incorporated an amperometric detector, operated in the oxidative mode, for the simultaneous determination of L-dopa and 3-OMD in human serum. A thin-layer cell, containing a glassy carbon working electrode held at +0.66 V versus Ag/AgCl, was used to monitor the eluent from a μBondapak C_{18} column; the eluent was 0.1-M ammonium phosphate pH 4.3. Michotte *et al.* [78] reported on a similar study but used a mobile phase containing MeOH-(0.1-M sodium acetate–20-mM citric acid) (1:99) with 1-mM octane 1-sulphonic acid, 0.1-mM EDTA, 1-mM butylamine adjusted to pH 3.0 with *o*-phosphoric acid; these authors found it beneficial to incorporate a clean-up step, using a sephadex G-10 column, prior to LCEC.

Baruzzi *et al.* [79] preferred to employ a coulometric detector for the simultaneous quantitation of L-dopa and 3-OMD. This consisted of a cell containign series dual porous graphite electrodes and was operated in the redox mode. The first electrode was held at +0.04 V and the second at −0.3 V at which the currents were measured; a conditioning electrode was also incorporated upstream of the dual cell and this electrode was held at +0.35 V. Chromatographic separation was carried with a Nucleosil C_{18} (5-μm) column, which was protected by a guard column. The mobile phase consisted of dihydrogen phosphate (0.05-M) sodium acetate (0.05-M) sodium dodecyl sulphate (0.0007-M), Na_2EDTA (0.002-M) adjusted to pH 3.1 with 2-M orthophosphoric acid; this was then mixed with acetonitrile in the ratio 87.5:12.5 v/v. A simple extraction procedure was used to separate the substances from plasma. Plasma (100 μl) and 1.2-M perchloric acid (100 μl) were diluted to 1.0 ml with HPLC-grade distilled water and vortex-mixed for 30 s; after centrifugation for 10 min at 2500 *g* a 30-μl aliquot was injected onto the column. The recovery for both compounds was 94–99% and addition of an internal standard was said to be unnecessary because only one sample preparation step was involved. The precision for L-dopa and 3-OMD was 2.8% (6.3 μM) and 5.2% (1 μM) respectively and the respective detection limits were 0.15 and 1.3 μM. These authors suggested that the guard column should be replaced after 60 injections.

Boomsma *et al.* [80] also reported on an LCEC method for the determination of L-dopa and 3-OMD. Separation was achieved at 35°C on a 3-μm Microsphere column; the mobile phase consisted of a phosphate buffer, dodecyl sulphate, EDTA and 25% methanol. The pH was 2.34, which was found to be critical for good chromatographic separation; the detector was set at +0.80 V versus Ag/AgCl. Fig. 3.12 shows the chromatograms obtained from a standard mixture and a plasma extract.

Interestingly, an LCEC method involving gradient elution has been reported by Ishimitsu and Hirose [81] that could measure L-dopa as well as catecholamines and *o*-methylated metabolites in human plasma. A combination of a reversed-phase column and an amperometric detector, contained a glassy carbon electrode, was

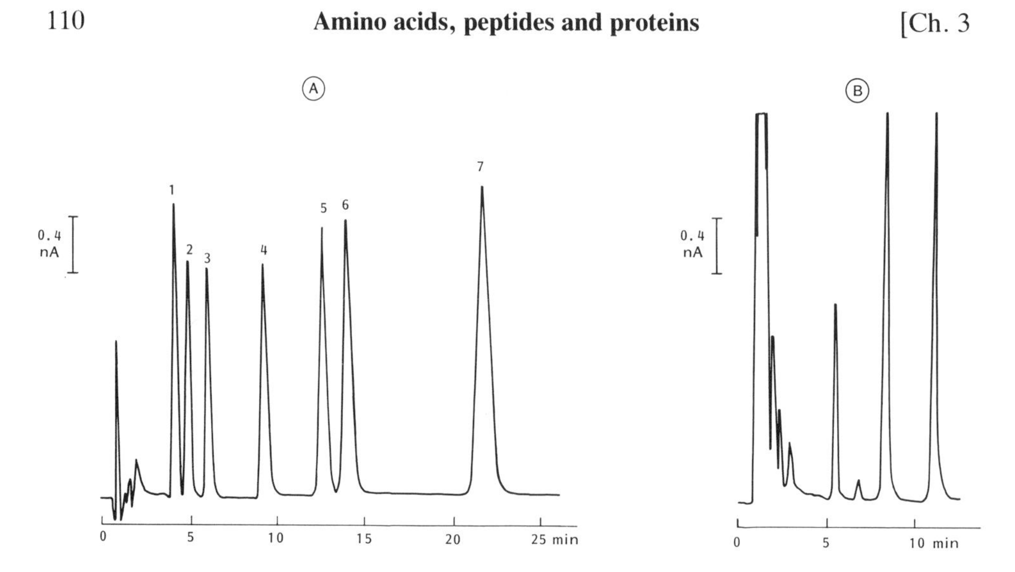

Fig. 3.12 — HPLC chromatogram (direct method). (a) Standard mixture of (1) noradrenaline (7.4 pmol), (2) adrenaline (6.8 pmol), (3) DOPA (6.3 pmol), (4) dopamine (8.2 pmol); (5) 3-OMD (23.7 pmol), (6) carbidopa (11.1 pmol) and (7) benserazide (19.5 pmol). (b) plasma sample containing 1.94 μmol l^{-1} of DOPA and 2.07 μmol l^{-1} of 3-OMD. (Reproduced from [80] by permission of the copyright holders.)

employed. Calibration was performed using an internal standard (*o*-tyrosine); linear plots were obtained over a range 0.25–100 ng ml^{-1}. The free L-dopa levels were found to be in the range 0.91–0.44 ng ml^{-1}. Peak identification was established by retention data and by co-injection; identification was further confirmed by peak height ratios taken at +0.3 V and +0.6 V. These authors reported that the developed LCEC method was more sensitive than a method using HPLC with fluorescence detection.

A method which could detect physiological concentrations of L-dopa in plasma and urine has been reported by Benedict and Risk [82]. In this method a minibore column (Brownlee, 2.1 mm I.D.) was employed which consisted of two cartridges; these were a 3 cm length of C_{18} and 22-cm length to C_8 (5-μm particle size) arranged in series. The mobile phase did not contain an organic modifier or an ion-pairing agent but comprised 50-mM citric acid–sodium acetate pH 2.35 with 0.2-mM EDTA. Amperometric detection was performed with a thin-layer cell containing a glassy carbon working electrode held at +0.7 V versus Ag/AgCl. Concentrations in plasma were stated to be 2.12±0.31 ng cm^{-3}; urine contained 59.6±9.7 ng ml^{-1} (pooled over 24 h).

Shum *et al.* [83] have also reported an LCEC method that was sensitive enough to measure physiological concentrations of L-dopa in plasma.

An interesting approach to the problem of resolving overlapping L-dopa and 6-hydroxydopa chromatographic peaks in the presence of a tyrosine peak has been described by Last [84]. Electrochemical detection was performed with a cell containing an array of 10 electrodes [84]; chromatographic traces could be recorded on each channel from a single injection. Fig. 3.13 illustrates the way in which peak removal could be used to obtain a clear L-dopa peak.

The upper trace corresponds to an applied potential of +0.570 V; the larger peak represents 105 ng of L-dopa and the smaller 90 ng of 6-hydroxydopa. The middle trace corresponds to +0.245 V at which L-dopa is not significantly oxidized. Because both traces are collected simultaneously, the middle trace can be subtracted from the upper trace to yield the difference chromatogram (L-dopa peak) shown in the lower trace. Although the column resolution is only 0.6, the voltammetric detector provided complete resolution. Furthermore, this technique allowed resolution and quantitation of both components from a single chromatographic injection. The detection limits were 1.7 ng and 11 ng for tyrosine and L-dopa respectively. The authors indicated that L-dopa pharmaceutical products may contain tyrosine and 6-hydroxydopa as impurities; hence the analytical application of the proposed method.

A new type of dual-electrode detection system for liquid chromatography, utilizing bare and film-coated glassy carbon electrodes in a parallel opposed configuration (Fig. 3.14), has been described by Hutchins *et al.* [85]; this was applied to the analysis of several tyrosine metabolites.

In this system, the coating consisted of cellulose acetate which had undergone hydrolysis [86]. It was shown that each of the analytes could be monitored simultaneously at both electrodes (Fig. 3.15); the peak ratio measured from these LCEC chromatograms was stated to be unique for each solute and, therefore, valuable for peak identification. To confirm the identity of an analyte, for example in a biological sample, it would be necessary to inject pure compounds under exactly the same conditions as the sample. Table 3.2 gives the calculated current ratios for tyrosine and some related compounds.

Amperometric detection, using d.c., normal and differential-pulse waveforms, was investigated by Dieker *et al.* [87] for the determination of several tyrosine metabolites. These authors concluded that the d.c. mode was the most favourable when no adsorption of oxidation products onto the electrode took place. When strong adsorption occurred, the normal pulse technique was more reliable but there was a slight loss in sensitivity.

The electrochemical detection of the tyrosine metabolites known as the catecholamines have recently been reviewed by Stulik and Pacakova [88,89]; therefore, they will not be further described here.

Tryptophan and its metabolites

Analytical methods for tryptophan have usually been designed to quantitate not only the parent amino acid but also its important metabolites.

An LCEC method for the determination of tryptophan, serotonin (XXVIII, 5-hydroxytryptamine) and 5-hydroxyindoleacetic acid (XXIX) in blood, urine, cerebrospinal fluid (CSF), brain and endocrine glands has been described by Koch and Kissinger [90]. These analytes were initially isolated from the samples on extraction columns (cation-exchange or sephadex). After elution, the appropriate fractions were collected and then analysed using two ODS chromatographic columns joined by stainless steel tubing. Two separate HPLC systems were employed. One

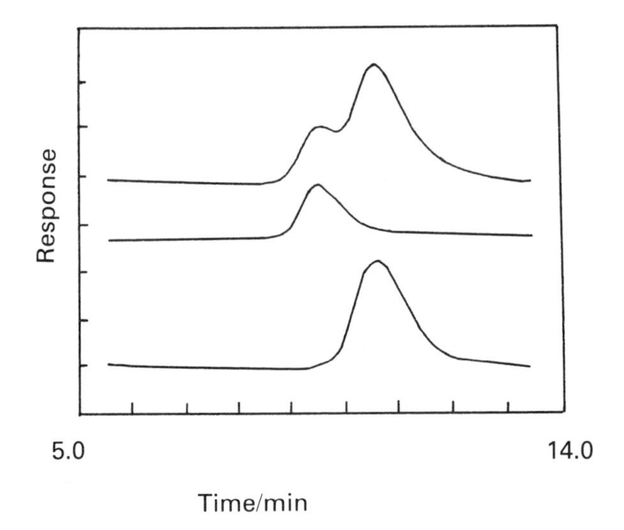

Fig. 3.13 — Difference chromatogram (lower trace) produced by numerical subtraction of the middle trace from the upper trace. Smaller peak is 6-hydroxydopa; larger peak is L-dopa. (Reproduced from [84] by permission of the copyright holders, Elsevier Science Publishers Physical Sciences and Engineering Division.).

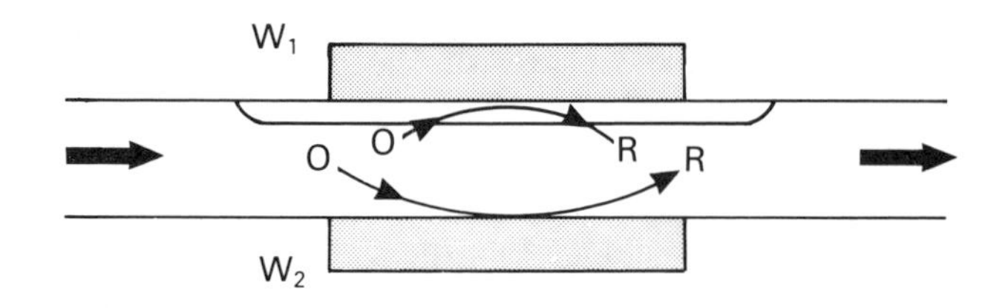

Fig. 3.14 — Schematic depiction of the parallel-opposed dual-electrode detector with bare and coated glassy carbon electrodes. (Reprinted with permission from L. D. Hutchins, J. Wang and P. Tuzhi, *Anal. Chem.*, 1986, **58**, 1019. Copyright 1986, American Chemical Society.).

employed a mobile phase containing 0.5-M ammonium acetate pH 5.1, with 15% methanol and the detector was set at +0.5 V; this was used to determine XXVIII and XXIX in separate injected fractions. The second system required a mobile phase comprising McIlvaine buffer pH 4.0 with 20% methanol, and a potential of +1.0 V was applied to the cell; this was used to determine tryptophan which eluted after 8 min. It was reported that the use of two such systems considerably reduced analysis times.

A similar investigation into the determination of tryptophan (VIII), 5-hydroxytryptamine(XXVIII), 5-hydroxyindoleacetic acid(XXIX) and 5-hydroxytryptophan (XXX) in plasma was described by Qureshi and Gokmen [91]. In this study, the column contained 5-μm ODS and the mobile phase consisted of 0.02-M trichloroace-

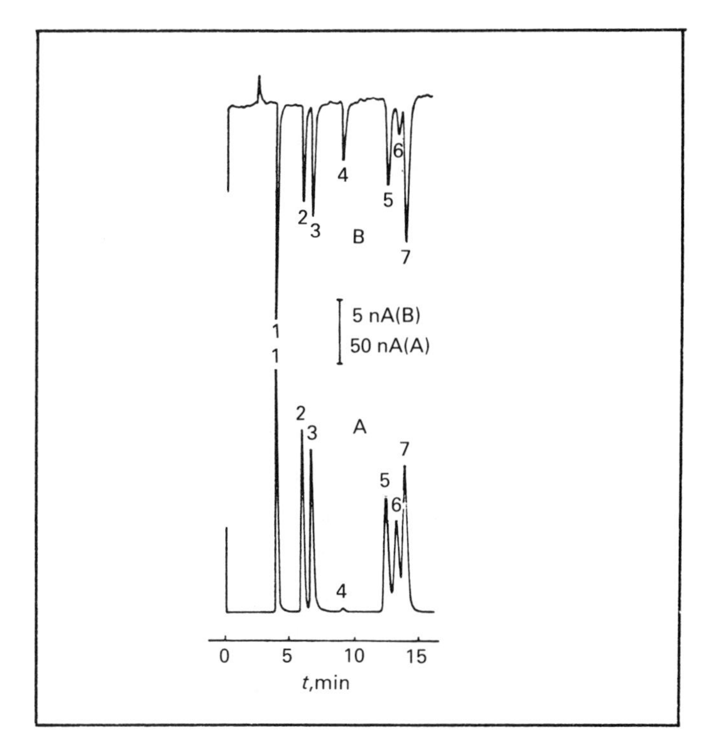

Fig. 3.15 — Chromatograms obtained simultaneously with dual-electrode operation: (a) bare electrode (b) coated electrode (30 min hydrolysis); operating potential (both electrodes), +0.90 V; flow rate 1.0 $cm^3 min^{-1}$, mobile phase 0.15-M chloroacetic acid solution adjusted to pH 3.5 with phosphoric acid (containing 0.2 g l^{-1} disodium EDTA); peak identities: (1) norepinephrine, (2) L-dopa; (3) epinephrine; (4) tyrosine; (5) dopamine; (6) α-methyldopa; (7) homogentisic acid; all present at the 100-ng level. (Reprinted with permission from L. D. Hutchins, J. Wang and P. Tuzhi, *Anal. Chem.*, 1986, **58**, 1019. Copyright 1986, American Chemical Society.).

Table 3.2 — Current ratios for parallel outputs for biogenic amines

Compound	mol. wt.	$i_{p,c}/i_{p,b}$
Norepinephrine	169	0.094
L-Dopa	197	0.056
Epinephrine	183	0.072
Tyrosine	181	1.300
Dopamine	153	0.077
α-Methyldopa	211	0.042
Homogentisic acid	168	0.100

(Reprinted with permission from L. D. Hutchins, J. Wang and P. Tuzhi, *Anal. Chem.*, 1986, **58**, 1019. Copyright 1986, American Chemical Society.).

tic acid–HCl pH 4.25, with 15% methanol and 1 mM EDTA; amperometric detection was carried out with a glassy carbon electrode held at +0.85 V; it was stated that the method was simple and reproducible.

A coulometric detection system was used in conjunction with a reversed-phase column for the LCEC determination of tryptophan (VIII) and catecholamines [92]. Operation of this detector in the screen mode offered increased selectivity for compounds which oxidized at high applied potentials. Fig. 3.16 shows the chromato-

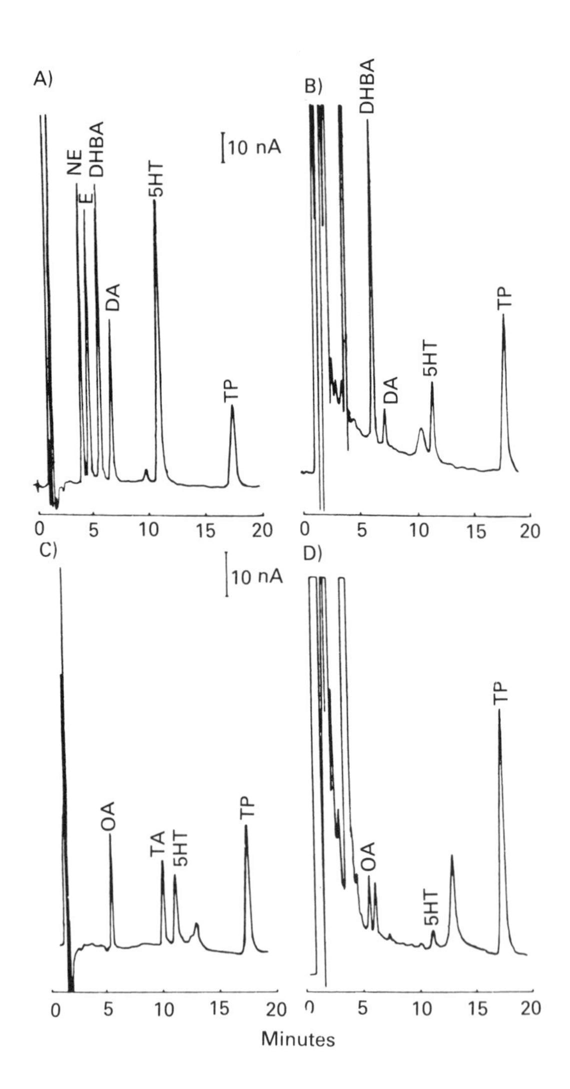

Fig. 3.16 — Chromatograms of 1 ng of standards (a,c) and cockroach nerve extracts (b,d) with detector 1 set at +0.5 V (a,b), detector 2 at +0.75 V (c,d). Flow rate 0.8 cm^3 min^{-1}. (The peak appearing at 13.1 min on detector 2 is an unidentified contaminant originating from the injector seal material). (Reproduced from [92] by permission of the copyright holders, Elsevier Science Publishers Physical Sciences and Engineering Division.).

grams obtained with this system; the advantage of this approach can be seen for octopamine (OA), which is free from interference when detector 2 is used; the potential interference DHBA is totally oxidized at detector 1 and is therefore 'screened out'. Stulik and Pacakova [88,89] have reviewed LCEC methods for tryptophan and its metabolites.

Neurotransmitter peptides
The electrochemical activity of peptides depends on the presence of common amino acids that are easy to oxidize, e.g. tyrosine and tryptophan.

Sauter and Frick [93] have reported that cholecystokinins can be determined in biological materials using LCEC. Cholecystokinin tetrapeptide and octapeptide sulphate were measured in rat brain cortex, hyppocampus, striatum and brain stem and the results were found to be comparable to those obtained by radioimmunoassay methods. The LCEC system used in this study was similar to those described in the previous two subsections; one slight difference was in the use of propanol as the organic modifier. Details on the methodology required in the assay were presented.

A thorough investigation of the experimental conditions required for the determination of some neuropeptides was described by Dawson *et al.* [94]. Again the LCEC system contained conventional components as described previously; the methods were applied to neural lobe arginine, vassopressin and striatal methionine enkephalin. This report [94] also contained a number of useful references concerning the LCEC determination of peptides; a brief mention of the LCEC determination of neuropeptides was also given in a review by Di Bussolo [95].

Glutathione
A very recent report described an LCEC assay for the simultaneous determination of reduced and oxidized glutathione using direct electrochemical oxidation of both compounds [96]. The authors used a cell containing dual porous graphite electrodes in the series configuration and operated in the 'screen mode'. The upstream electrode was held at +0.7 V and the downstream electrode (detector) at +0.9 V; in addition, a guard cell held at +0.95 V was placed between the pump and the injector to remove possible contaminants in the mobile phase. Current measurements were made only at the downstream electrode. The authors reported detection limits that were comparable with other electrochemical detection systems; these were 1.5 and 3.0 ng for GSH and GSSG respectively. This method was applied to the determination of the two species in small samples, such as a single mouse embryo weighing about 1 mg. One drawback appears to be loss in sensitivity; when 100 samples had been injected the signal for GSSG had reduced by 50%; therefore, frequent cleaning of the cell was required.

3.2.2.2 LCEC methods using chemically modified carbon electrodes

Sulphur-containing amino acids and related compounds
The sulphur-containing amino acids, glutathione and related compounds have been investigated in mixtures with each other, so it is convenient to discuss these compounds together.

As mentioned earlier (Section 3.2.1.7), carbon electrodes that have been modified with CoPC have the advantage that thiol-containing compounds may be detected at modest anodic potentials. This electrocatalyst, which can be incorporated in several different ways, has also been investigated by several different groups as a sensor for LCEC.

Halbert and Baldwin [70] investigated the possibility of measuring several sulphur-containing amino acids using a CoPC chemically modified electrode. These workers constructed a working electrode by mixing CoPC into the graphite powder/nujol oil matrix used to fabricate conventional carbon paste electrodes. HDVs were constructed from the response of the electrode, which was incorporated into a conventional thin-layer cell; however, a short length of narrow stainless steel tubing was inserted in place of the column between the injector and detector. The mobile phase consisted of 5% methanol–95% 0.05-M phosphate buffer pH 2.4 containing 0.001-M sodium octane sulphonate. The HDVs for cysteine (IX), homocysteine (XXII) and *N*-acetylcysteine (XXIII) all showed an anodic wave that reached a maximum between +0.7 and +0.8 V versus Ag/AgCl and rapidly decreased at higher potentials (Fig. 3.17). For LCEC an applied potential of +0.75 V was selected for detection, following separation of the three compounds on a octadecylsilane column. Calibration curves for each of the amino acids were linear over the range 2.7 to 270 pmol injected; the detection limit corresponded to the former value. The precision was examined by making repeat injections of a solution containing 1×10^{-4}-M cysteine, with the electrode being replaced between each injection. The authors reported that the relative standard deviation was 6.98% (n=9) and that this was not appreciably less than that expected for the repeated renewal of ordinary carbon paste surfaces.

In a subsequent report, Halbert and Baldwin [97] applied the LCEC method, using a CoPC modified electrode, to the determination of cysteine and glutathione in whole blood and plasma. The samples were treated with a mixture containing EDTA and *o*-phosphoric acid; after thorough mixing a 100-μl aliquot of the suspension was injected through a nylon filter onto the chromatographic column. The resulting chromatograms for whole blood showed complete separation of GSH and cysteine from each other, as well as from other naturally occurring substances; the glutathione levels were found to be in the range 0.74 to 1.59 mM (n=5). The plasma cysteine levels were reported to be in the range 1.83 to 10.36 μM (n=5). In this study, it was found that plasma contained little or no glutathione. One drawback with the CoPC/carbon paste electrode was that the response decreased to about 75–80% of the initial current when sequential injections of 20 or more blood samples were made. However, it was pointed out that the electrode surface was easily renewable in less than 10 min.

The present author and coworkers have also been interested in the application of chemically modified electrodes, in particular to determine reduced glutathione in biological fluids. Systematic studies were carried out [72] on carbon paste electrodes modified with CoPC in a similar manner to Halbert and Baldwin [70,97]. Initial studies were performed with a thin-layer cell together with a mobile phase consisting of 0.1-M phosphate buffer (pH 5.0); these were used in a flow injection system to study the electrochemical characteristics of GSH. In our studies we also found that the response for GSH decreased with time, and like the previously mentioned workers [70,97], we also concluded that the CoPC probably leached from the

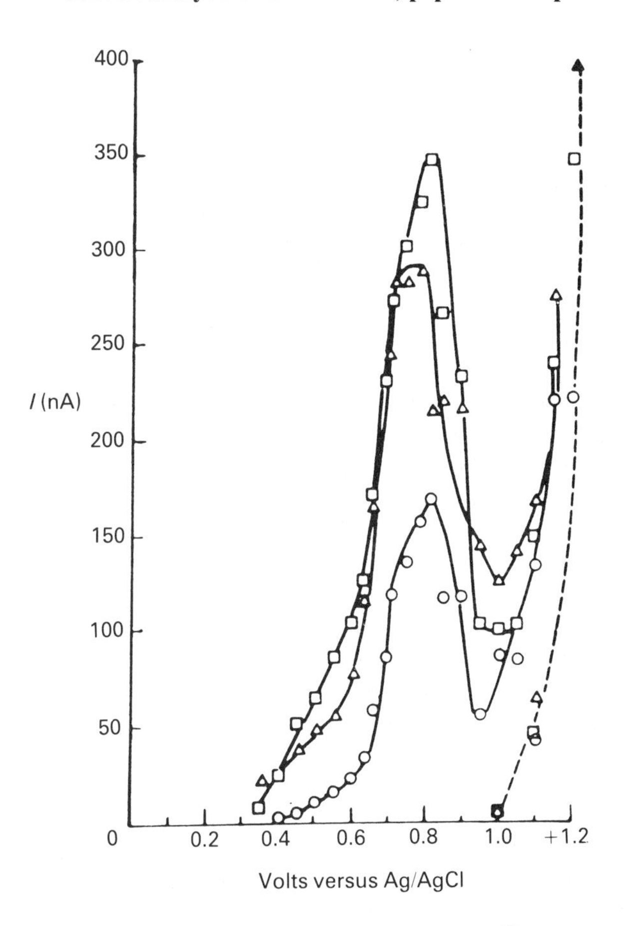

Fig. 3.17 — Hydrodynamic voltammograms of cysteine (○), homocysteine (□), and *N*-acetylcysteine (△) at a 2% CoPC electrode. Mobile phase was 5% CH_3OH/95% 0.05-M phosphate buffer (pH 2.43) with 1.0-mM octane sulphonate. Flow rate was 1.0 cm^3 min^{-1}. (Reprinted with permission from M. K. Halbert and R. P. Baldwin, *Anal. Chem.*, 1985, **57**, 591. Copyright 1985, American Chemical Society.).

electrode during analysis. Attempts were made by us to improve the stability of the CoPC/carbon paste electrode; for example, the pH of mobile phase was varied, different compositions and also different types of pasting agent were used, MeOH was added to the mobile phase, and the flow rate was reduced. The applied potential was also investigated. However, none of these parameters appeared to improve the stability of the electrode; therefore we decided to investigate an alternative approach.

Our initial studies were performed with an electrode which had been fabricated by mixing epoxy resin with a 5% w/w mixture of CoPC in graphite powder, in the ratio 1:1. This was packed into a cell base of our own design and cured for a measured time; the resulting hardened surface was polished with wet and dry emery paper (1200 grade) and a final polish was performed with a slurry prepared from

0.03-μm aluminium oxide. Fig. 3.18 shows the cell base containing the CoPC modified carbon electrode.

The cell bottom was attached to a standard cell top with a 50-μm spacer in between; a reference cell compartment containing an SCE was connected to the cell top and a stainless steel outlet formed the counter electrode (Fig. 3.19).

We first investigated this electrode as a detector with FIA. Repeat injections of 15 ng of GSH were made over a period of 270 min without any loss in sensitivity; the coefficient of variation of the currents was 3.1% (n=22). These results indicated that the developed CoPC/epoxy/graphite electrode could prove to be an extremely powerful sensor for LCEC determinations of GSH. Our early chromatographic studies have been carried out with an ODS (μBondapak) column and a mobile phase consisting of 0.05 M phosphate buffer pH 3.0 containing 0.1% w/w EDTA; the detector potential was set at +0.45 V versus SCE. A linear calibration graph was obtained over the range 0.24 to 30.7 ng injected; the detection limit was calculated to be 450 pg GSH when the full-scale deflection was 5 nA. The method has been applied to human serum samples spiked with GSH [98]. After addition of EDTA and phosphoric acid the samples were filtered and injected onto the column. The recovery was determined from the resulting chromatograms and was found to be greater than 90%. We have applied this method to plasma samples taken from several healthy adults and Fig. 3.20 shows an LCEC chromatogram that was obtained from one of these subjects.

The calculated GSH levels agreed with literature values, which suggests that the method is likely to be suitable for clinical studies; the detection limit also suggests that this could also be applicable to the measurement of depressed circulating plasma levels.

While we were preparing our reports [72,98] on the development of a reusable CoPC sensor for GSH a similar electrode material was briefly reported to be suitable for cysteine determination [99], using FIA. Although no real samples had been analysed, it is probable that a suitable reversed-phase column could be employed to determine cysteine in biological fluids by LCEC.

3.2.2.3 LCEC methods with copper working electrodes

Several studies using potentiometric methods indicated that it was possible to detect amino acids through their complexation reactions with copper ions [100–102]. It was quickly realized by Kok *et al.* [103] that this basic principle could also be applied in amperometric devices and that such devices should have the advantage of linearity of response over a much wider concentration range. From the results of their investigations these workers [103] suggested that when an amperometric detector contains a metallic copper working electrode, which is poised at slightly positive potentials, a film of hydrated oxides is formed on its surface; this produces a small constant anodic current. In the presence of substances that form stable complexes with cupric ions, e.g. amino acids, the solubility of the oxide is enhanced and therefore the anodic current increases.

In a subsequent study Kok *et al.* [104] investigated the possibility of using the copper amperometric sensor as a detector in liquid chromatography. The separation of amino acid mixtures was carried out with reversed-phase systems (Lichrosorb RP-18, 5 μm) and polymer type material (PRP-1); both phosphate and carbonate

Fig. 3.18 — Flow-through cell bottom showing CoPC chemically modified carbon-epoxy working electrode.

buffers were found to be suitable but not borate buffer. The eluents from these columns were monitored by a conventional flow-through cell containing a 3-mm-diameter copper disc electrode; a platinum auxiliary electrode and an Ag/AgCl–1-M LiCl in methanol/water (1:1) reference electrode were also incorporated into the cell. The potential of the working electrode was held at +0.1 V. The copper electrode was polished regularly and rinsed with 5-N HNO_3. Before a series of experiments the electrode potential was set at −0.3 V for 5 min and at +0.1 V for at least 15 min. It was shown that while threonine (VI) was not retained, valine (III), methionine (XXXI), isoleucine (V), leucine (IV), phenylalanine (XI) and tryptophan (VIII) could be resolved and determined in about 12 min. The linear calibration range for these amino acids was 3×10^{-6} to 1×10^{-3} M. The detection limits for a conventional column ranged from 2.4 to 10.3 ng; with a microbore column the range was from 0.4 to 1.2 ng. The microbore column was then successfully applied to the determination of phenyalalanine and tryptophan in urine samples (Fig. 3.21).

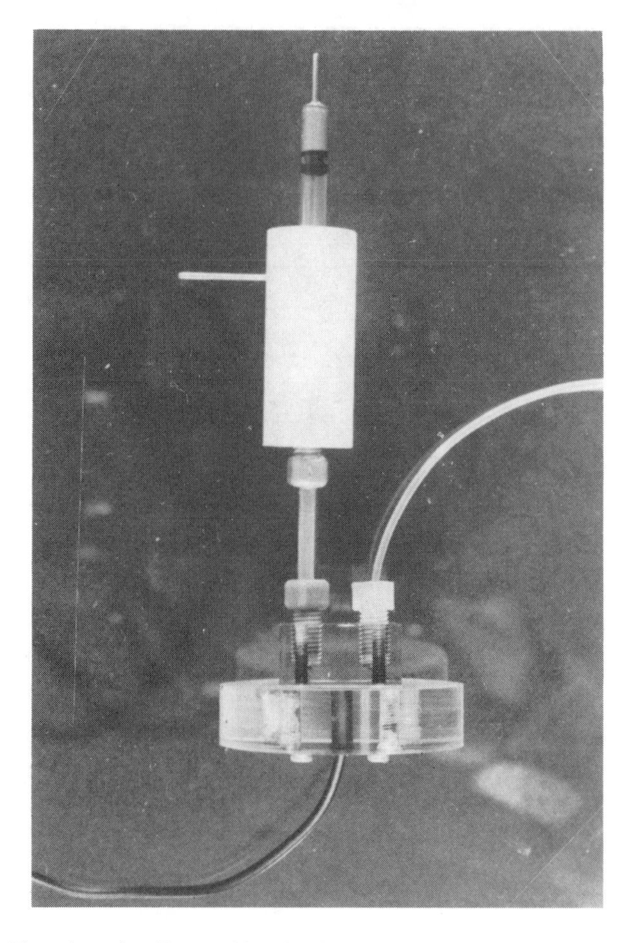

Fig. 3.19 — Flow-through cell comprising the CoPC chemically modified graphite electrode, together with reference compartment containing stainless steel auxiliary electrode and Ag/AgCl reference electrode.

This same group [105] later described the application of a conventional reversed-phase column, with the same copper working electrode, for the determination of phenylalanine in urine. This particular study was concerned with screening for phenylketonuria, and the authors compared the results from LCEC with those from an amino acid analyser. The results obtained with the two methods were in reasonable agreement; the calculated correlation coefficient was reported to be 0.991. It was stated that the LCEC method had several advantages: relatively short elution time (less than 15 min), simplicity and low cost of the apparatus used.

Stulik *et al.* [106] have developed two detection systems for amino acids which utilize copper working electrodes; both of these are different to that described by Kok *et al.* [104,105]. In one of these [106] a tubular copper electrode was constructed from a copper rod so that the diameter was 0.5 mm and the length 3 mm; this was screwed directly onto the metallic mantle of a glass microcolumn. The internal

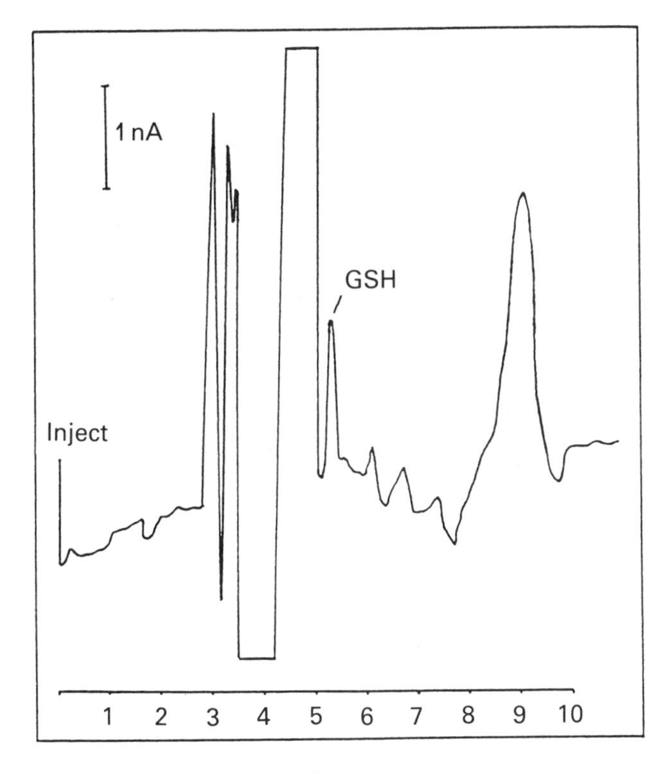

Fig. 3.20 — LCEC chromatogram of a normal human plasma sample. Final concentration was 1.87 μM. (Reproduced from [98] by permission of the copyright holders. Royal Society of Chemistry.).

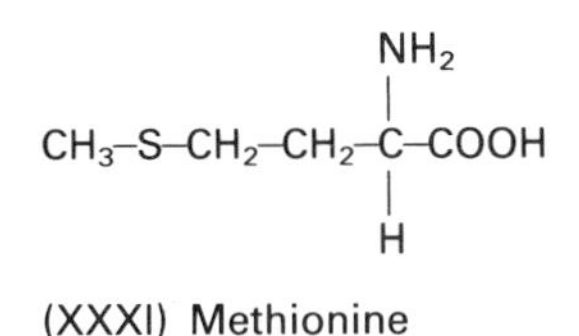

volume of this particular cell was 0.6 μl. The other detector comprised strands of copper wire (0.08–0.1 mm) as the working electrode and was based on the previous design [107]; the cell volume was 0.2 μl. In order to activate these detectors it was necessary to first carry out polarization in a similar manner to that described earlier [104]. Separation of the various amino acids was carried out on reversed-phase chromatographic columns; either a glass column, 150×1 mm i.d. containing Separon 5 μM, or a steel column, 250×4.6 mm i.d. containing Partisil 10 μm was used. In this study several mobile phases were investigated. The first contained 0.025-M aqueous

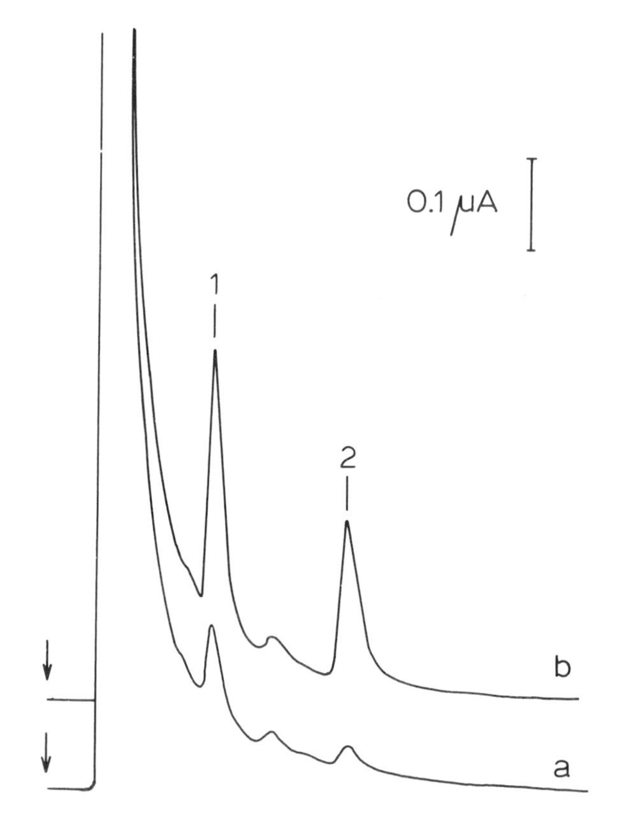

Fig. 3.21 — Determination of phenylalanine and tryptophan in urine. Column: 250×1.1 mm i.d., 7.7-μm Lichrosorb RP-18. Mobile phase: 0.025-M phosphate buffer, 10% methanol, pH 7.2. Urine diluted 1:2: (a) blank; (b) spiked with 10^{-4}-/M phenylalanine (1) and tryptophan (2). Reproduced from [104] by permission of the copyright holders, Elsevier Science Publishers, Physical Sciences and Engineering Division.).

KH_2PO_4–MeOH pH 6.8; the other mobile phases consisted of different concentrations of aqueous ammonia (0.1, 0.5 and 1.0 M). For calibration purposes the former was used at a flow-rate of 0.3 cm^{-3} min^{-1}, and this was used in conjunction with the copper strand electrode held at +0.15 V; the calibration parameters for the various amino acids are summarized in Table 3.3. The method was successfully applied to the determination of lysine, methionine and choline in fodder biofactors.

Further studies, using a copper tubular electrode, have been carried out by Stulik *et al.* [108]. In this investigation the authors carried out the determination on mixtures of dipeptides using a glass column (15×1 mm) containing Separon SIC 18 (7 μm); the mobile phase consisted of 0.025 M KH_2PO_4–10% methanol. The workers reported detection limits between 5 and 250 ng.

A method has also been described by Ichimura *et al.* [109] for the determination of amino acids using LCEC with a copper working electrode.

Table 3.3 — Calibration curve parameters

Amino acid	Correlation coefficient	Slope of the regression line (αA/μg)	Detection limit (ng)[a]
Tryptophan	0.9985	8.2	8.0
Valine	0.9961	11.8	4.1
Phenylalanine	0.9969	12.6	1.9
Tyrosine	0.9996	7.6	4.9
Alanine	0.9975	6.9	9.0
Methionine	0.9999	10.6	4.5
Proline	0.9930	2.5	18.0
Isoleucine	0.9974	5.3	12.8
Aspartic acid	0.9887	5.4	13.0
Glycine	0.9999	16.4	1.7
Lysine	0.9997	9.6	6.7
Glutamine	0.9988	9.6	4.8
Aspartate	0.9992	5.7	10.5
Histidine	0.9989	57.4	0.4
Serine	0.9938	52.3	0.5
Arginine	0.9997	36.2	0.8
Cysteine	0.9804	17.7	1.6
Threonine	0.9930	29.9	1.1
Glutamic acid	0.9997	54.8	0.6
Leucine	0.9959	17.5	1.6

[a]The detection limit is twice the peak-to-peak noise.
Column 2, detector 1. Mobile phase: 0.025 M KH_2PO_4+10% methanol; flow-rate 0.3 ml min^{-1}. Applied potential: +0.15 V. Samples 20 μl.
(Reproduced from [106] by permission of the copyright holders, Elsevier Science Publishers Physical Sciences and Engineering Division.)

3.2.2.4 LCEC methods with mercury, gold amalgam and gold electrodes

Sulphur-containing amino acids and related compounds

In earlier sections it has been shown that amino acids, peptides and related compounds containing the thiol moiety can be electrochemically detected; the reactions are believed to involve the oxidation of mercury in the presence of the thiol. These electro-oxidative processes have been exploited for the determination of several amino acids and related compounds. In this subsection glutathione is considered with the amino acids since this will prevent repetition.

Rabenstein and Saetre [110] were among the first to apply an LCEC method to the determination of sulphur-containing compounds. In this investigation they used a flow-through cell comprising a mercury pool working electrode, a platinum auxiliary electrode and an SCE reference electrode; the detector potential was held at 0.0 V. Chromatographic separations were performed with a Zipax SCX cation-exchange resin and a mobile phase consisting of phosphate–citrate buffer pH 2.5. With this system it was possible to separate and detect a mixture containing *N*-acetylcysteine, GSH, cysteine and homocysteine; these compounds were eluted in the order given. In the same study, the authors also investigated the amino acids glycine, lysine, aspartic acid, glutamic acid and histidine but found that none of these gave a response under the conditions used.

Further studies on the determination of cysteine were also reported by Saetre and

Rabenstein [111]. These same authors have applied the technique to samples containing both thiol and disulphides [112]. Quantitation of the disulphides in physiological fluids was performed through a two-stage process involving first exhaustive electrolysis to reduce disulphides to thiols, followed by quantitation of the thiols by LCEC. The concentration of the thiols was also determined prior to electrolysis and the difference was taken to measure the disulphides. This procedure has the drawback that a considerable analysis time is required; there is also the possibility of errors arising owing to the formation of new compounds during electrolysis.

An on-line reduction of the disulphide amino acid cystine has been described by Eggli and Asper [113]. These authors used a column containing amalgamated silver granules to generate the thiol, which was then detected downstream at a mercury pool electrode. Though this method demonstrated the utility of an on-line approach there was a problem with band-broadening.

Allison [114] has used a similar concept to this but has eliminated the problem of band-broadening by the use of dual electrodes. In this detection system two mercury/gold (gold amalgam) electrodes are used in a series arrangement with reduction of the disulphide to thiol at the upstream electrode, followed by thiol detection downstream (Fig. 3.22).

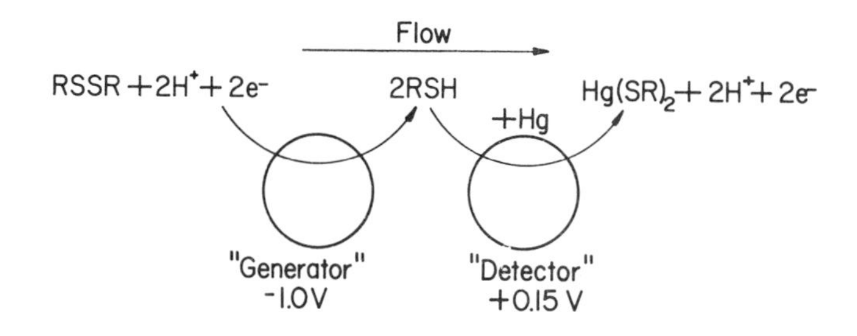

Fig. 3.22 — Schematic diagram of dual Hg/Au detector and reactions that occur at each electrode. (Reproduced from [114] by permission of the copyright holders, Bioanalytical Systems Inc.).

The upstream electrode behaves as a novel on-line post-column reactor of negligible dead volume, making the dual electrode detector suitable for liquid chromatographic separations. The authors applied this to the separation and determination of a mixture containing cysteine and cystine, together with the oxidized and reduced forms of glutathione. This same system was later applied to the determination of GSH and GSSG in whole blood (Fig. 3.23) [115].

In this study, whole blood was analysed after the addition of Na_2EDTA and $HClO_4$, followed by centrifugation and filtration; the concentration of GSH and GSSG were found to be 1.1 mM and 1.8 μM. Since the respective detection limits were reported to be 3.5 pmol and 5.7 pmol, this would indicate that the method could easily be applied to subnormal circulating levels.

A very similar approach to the determination of glutathione, cysteine and other

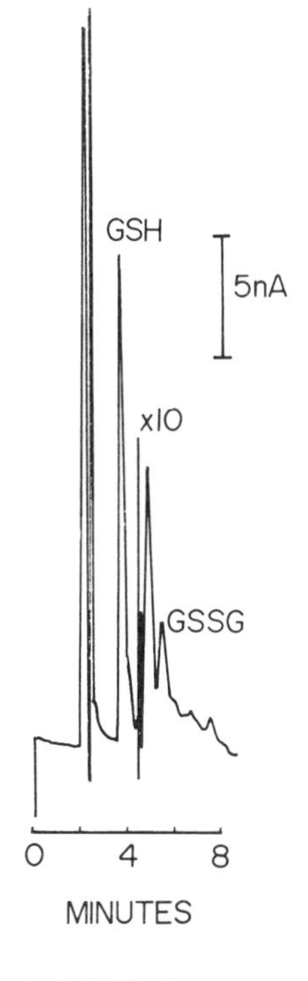

Fig. 3.23 — Downstream electrode (+0.15 V) chromatogram of whole blood filtrate with upstream electrode set at −1.0 V. Mobile phase, 99% 0.1-M nonochloacetate, pH 3.0/1% methanol. (Reprinted with permission from L. A. Allison and R. E. Shoup, *Anal. Chem.*, 1983, **55**, 8. Copyright 1983, American Chemical Society.).

thiols has recently been described by Richie and Lang [116]. The authors applied the method to human blood, rat liver and hippocampus, mosquito and spinach leaf. A conventional reversed-phase column was used and an ion pairing agent was incorporated into the mobile phase. Samples could be analysed in 15 min and the calibration graphs were rectilinear from 5 to 1600 pmol.

The thiols penicillamine and *N*-acetylcysteine are of considerable therapeutic importance; the former has been used in the treatment of Wilson's disease, cystinuria and rheumatoid arthritis [117,118] and the latter is used as a mucolytic drug in the treatment of bronchitis [119]. It was reported by Drummer *et al.* [120] that electrochemical detection with mercury/gold electrodes has become the method of choice for penicillamine; this is due to the high selectivity and ease of operation of the detector, and because little sample preparation is required. These authors [120] described a method for the determination of the two thiols in which a reversed-phase column was used to separate the species from blood samples, which had previously been deproteinated with perchloric acid. The mobile phase contained 4% methanol in 0.1% (v/v) phosphoric acid and 10-mM sodium sulphate; this was continually purged with nitrogen during analysis. The eluant from the column was monitored at +0.17 V versus Ag/AgCl. The chromatographic peaks for both drugs were well-resolved and it was also possible to monitor GSH simultaneously. In this study, it was found that *N*-acetylcysteine could be used as an internal standard for penicillamine when the former was not being administered to the patient, and vice versa.

Other reports have also appeared on the application of gold/mercury electrodes for the determination of penicillamine [121,122].

This same drug has been determined by means of its electro-oxidation at a simple gold electrode [123]. The drug was separated from deproteinated urine, or plasma, on a cation exchange column; for plasma, the mobile phase consisted of 0.02-M diammonium hydrogen citrate adjusted to pH 2.2 with phosphoric acid, and for urine the same buffer was used but the concentration was reduced from 0.02 M to 0.01 M of the same solution. The eluent from the column was monitored at a potential of +0.8 V versus Ag/AgCl. In this study, the authors reported that the sensitivity of the electrode decreased slightly with time, but this could be accounted for by a suitable injection sequence. The method has been automated for the determination in the biological fluids described.

3.2.2.5 LCEC methods involving derivatization procedures

As is apparent from previous sections of this chapter, the majority of naturally occurring amino acids are not easily oxidizable or reducible at conventional electrodes; therefore, procedures have been investigated for the purpose of preparing suitable electroactive derivatives for different classes of amino acids. In this section, the application of LCEC to the determination of these amino acid derivatives will be discussed.

o-Phthaldehyde derivatives

Recently, Joseph and Davies [124] reported on the application of the orthophthalaldehyde/beta-mercaptoethanol (OPA/BME) reaction with a variety of amino acids produce compounds that were electro-oxidizable; the basic reaction is that shown in Fig. 3.24 [125].

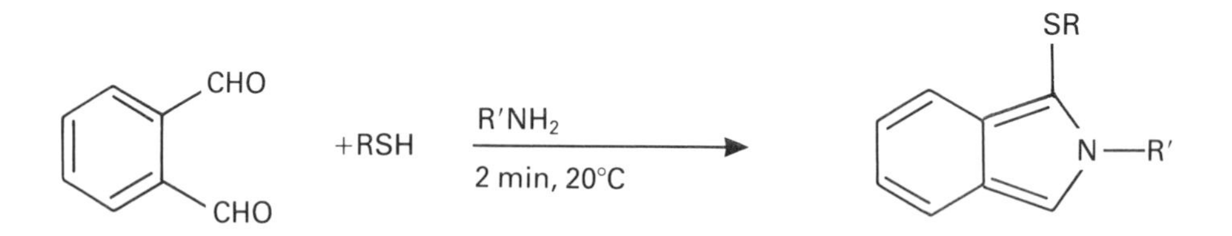

Fig. 3.24 — *o*-Phthalaldehyde/β-mercaptoethanol (OPA/BME) derivatives of amino acids. Reproduced from [125] by permission of the copyright holders, International Laboratory.).

The procedure reported by these authors [124] was a modification of that described by Lindroth and Mopper [126]. This involved the dissolution of 27 mg of *o*-phthalaldehyde in 500 μl of absolute ethanol, and addition of 5 ml of 0.1-M sodium tetraborate followed by 20 μl of mercaptoethanol. The reagent was kept overnight before use and 10 μl mercaptoethanol added if required to maintain maximal yield. For plasma analysis a 50-μl aliquot of sample was deproteinized with 4 volumes of methanol by thorough mixing and standing for 10 min at 4°C and centrifuged for 5 min. Water (50 μl) or mixtures of appropriate standard amino acid solutions were carried through the sample procedure. One volume of the supernatant was reacted with 4 volumes of the OPA reagent at room temperature. It was necessary to inject the samples at timed intervals of 2 min after mixing and 20-μl aliquots were injected. The authors investigated both isocratic and gradient elution methods for separation on a reversed-phase column; with the latter, 15 amino acids could be detected in human plasma (Fig. 3.25).

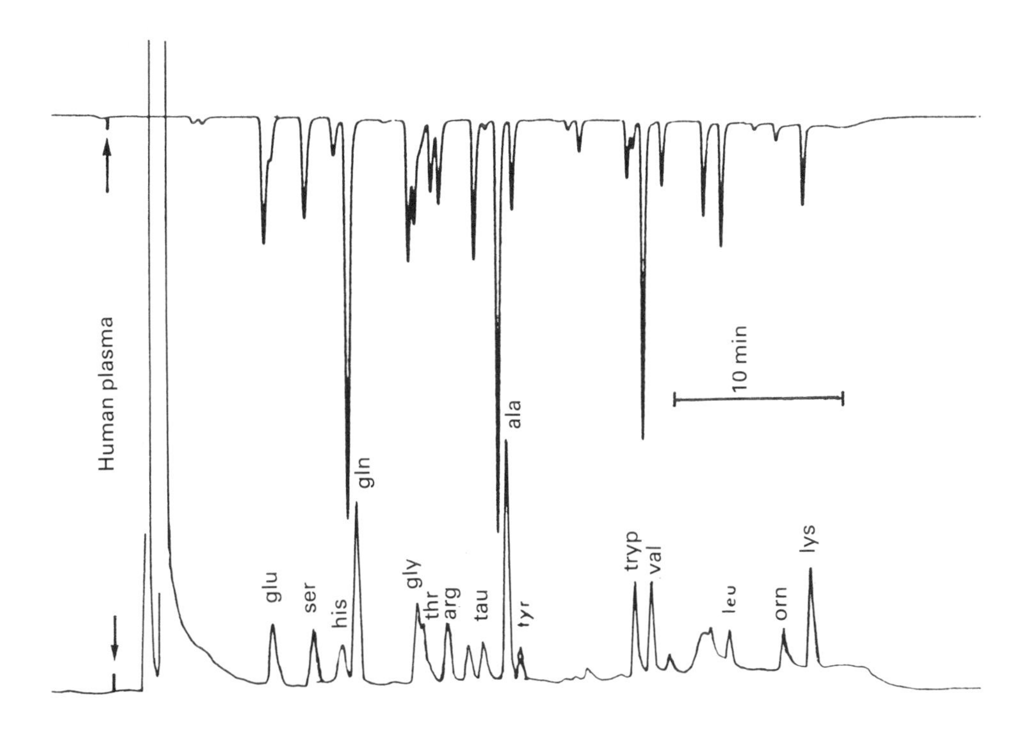

Fig. 3.25 — Gradient elution of amino acids in human plasma. Lower trace electrochemical detection at +0.5 V, fsd=50 nA; first peak is due to reagent. Upper trace fluorimetric detection, fsd 10 mV, ×1 setting. (Reproduced from [124] by permission of the copyright holders, Elsevier Science Publishers Physical Sciences and Engineering Division.).

The gradient involved two eluents and consisted of 0.05-M sodium phosphate (pH 5.5)/methanol, 80:20 for A, and 20:80 for B. A gradient of 0 to 10% B over 10 min, then to 85% B over 30 min, followed by 85 to 0% B in 5 min and 10-min re-equilibration was employed. This gradient elution was reported to be possible only if the level of metal ions present was reduced by the use of sintered glass instead of metal frits on the solvent inlet lines. The LCEC method was applied to other biological fluids, including human gastric juice, human CSF, and rat brain.

Allison *et al.* [127] have attempted to decrease the separation time on the chromatographic column, following the OPA derivatization procedure, for the LCEC determination of a variety of amino acids. In this investigation, short chromatographic columns (100×4.6 mm) containing a 3-μm ODS stationary phase were employed and it was shown that the derivatives of 20 amino acids couls be separated in under 10 min. The detection limit was about 500 fmol when the mobile phase contained 0.05-M $NaClO_4$–0.005-M sodium citrate pH 5.0/tetrahydrofuran/methanol; the detector potential was set at +0.7 V versus Ag/AgCl. This method was applied to the determination of amino acids in commercially available beer samples; such studies could be important in understanding yeast performance, effect of barley variety and also the development of new malting varieties [4]. Fig. 3.26 shows a typical chromatogram of a beer sample following the derivatization procedure; amino acid concentrations ranged from 33 μM (glutamine) to 1480 μM (alanine).

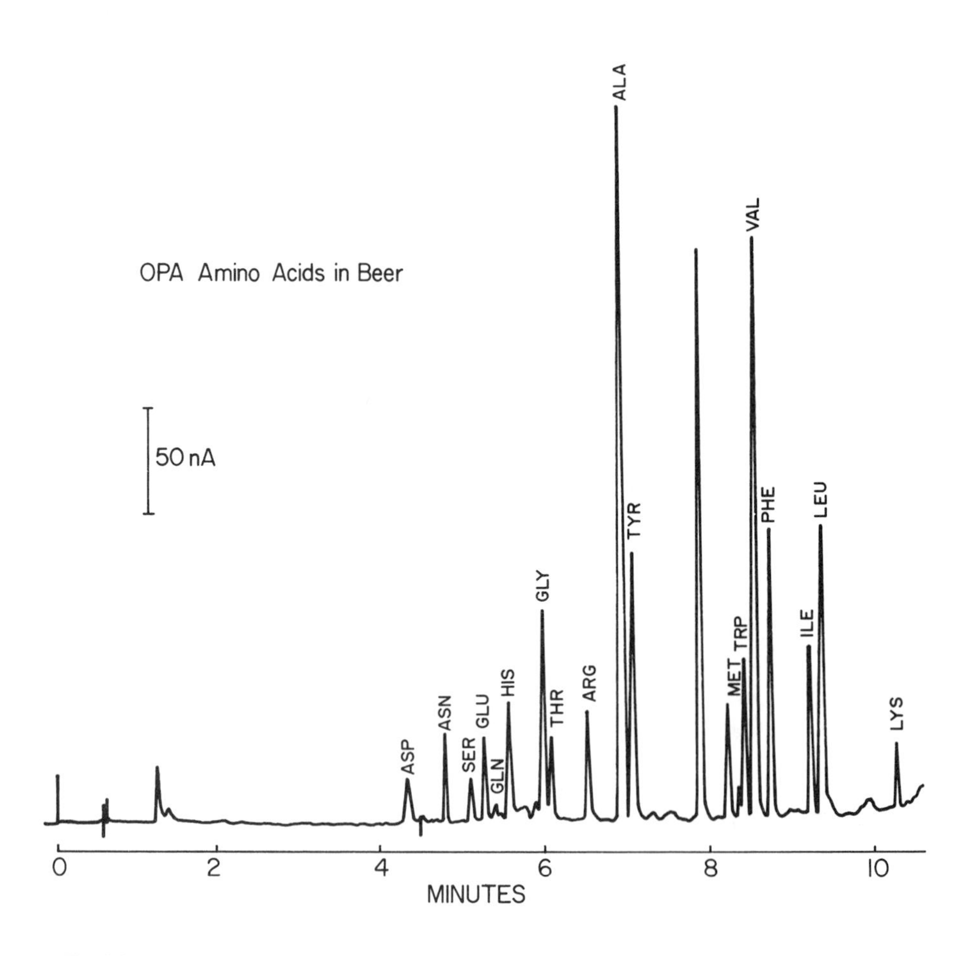

Fig. 3.26 — Chromatogram of OPA/βMCE derivatives of the amino acids in a commercial beer sample. (Reprinted with permission from L. A. Allison, G. S. Mayer and R. E. Shoup, *Anal. Chem.*, 1984, **56**, 1089. Copyright 1984, American Chemical Society.).

The primary drawback with this procedure was the lack of stability of the OPA/BME adducts. In particular, those amino acids with small side-chains, such as glycine, taurine and 4-aminobutyric acid, formed the least stable derivatives; it was considered likely that a more bulky thiol would be more suitable than mercaptoethanol in the reaction. In an attempt to improve the stability of OPA derivatives a study was carried out on eight different thiols with GABA (XXXII) and methionine;

$$HOOC{-}CH_2{-}CH_2{-}CH_2{-}NH_2$$

(XXXII) γ-Aminobutyric acid

of these *t*-butylthiol (*t*-BuSH) appeared to be most promising. The electrochemical behaviour of the *t*-BuSH derivatives of different classes of amino acids was studied; as shown in Fig. 3.27 the substitution did not affect the electrochemical behaviour of the resulting isoindole derivative.

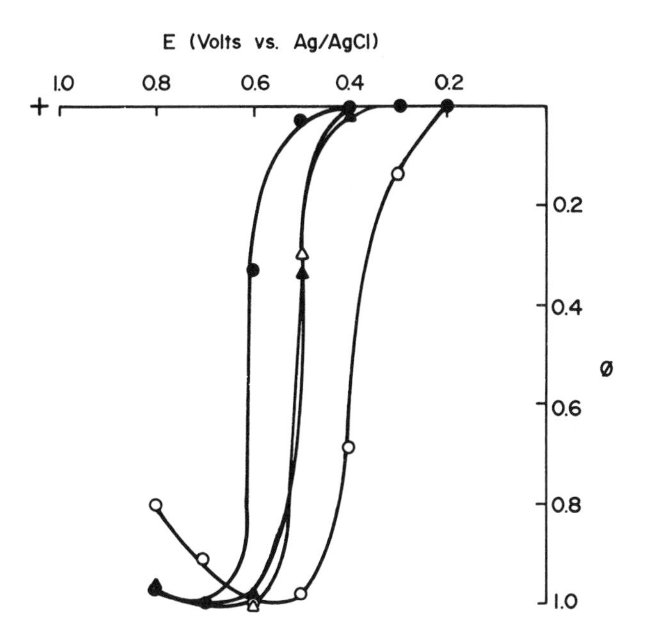

Fig. 3.27 — Hydrodynamic voltammograms for *o*-phthalaldehyde/*tert*-butylthiol derivatives of γ-aminobutyric acid (●), methionine (▲), tryptophan (○), and leucine (△) in a mobile phase consisting of 60% 0.05-M sodium perchlorate/0.005-M trisodium citrate buffer, pH 5.0 with 40% acetonitrile, ϕ is defined as the ratio of the peak curent to the limiting current. (Reprinted with permission from L. A. Allison, G. S. Mayer and R. E. Shoup, *Anal Chem.*, 1984, **56**, 1089. Copyright 1984, American Chemical Society.).

In addition, these compounds were found to possess half-lives of several hours. It was suggested that the method could also be used for the determination of important neurotransmitter amino acids in the rat brain; glutamine, taurine, aspartic acid, glutamic acid, glycine and GABA could be separated and detected in about 8 min by gradient-elution LCEC.

Zielke [128] has also investigated the derivitization procedure using *t*-BuSH for the determination of amino acids in brain samples; however, in this case chromatographic separations were achieved with only one pump. Fig. 3.28 shows the chromatogram obtained for an extract of rat cerebellum using a thin-layer cell

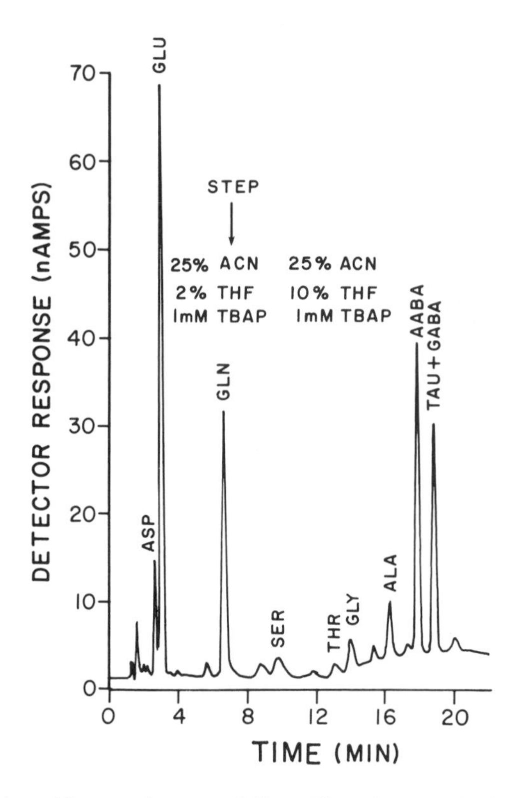

Fig. 3.28 — Amino acid pattern in rat cerebellum. The column was initially equilibriated with 50-mM potassium phosphate, pH 7.0, 1-mM EDTA, 25% acetonitrile, 2% THF and 1-mM TBAP. At 7 min the mobile phase was changed to include 10% THF. The volume injected corresponded to 330 ng of tissue, wet weight. Reproduced from [128] by permission of the copyright holders, Elsevier Science Publishers Physical, Sciences and Engineering Division.).

containing a glassy carbon electrode held at +0.7 V versus Ag/AgCl. The inclusion of the ion-pair reagent TBAP was necessary to separate aspartate and glutamate.

Hikal *et al.* [129] have also carried out studies on the determination of various amino acids in different regions of the rat brain. An LCEC method, using OPA derivatization, was applied to studies into the effects of tetrahydrocannabinol (THC) and trimethyl tin (TMT) compounds. Other workers have been interested in the application of OPA derivatization, mainly to measure sulphur-containing amino acids and related compounds [130–132].

Hoskins *et al.* [133] have performed systematic studies to try and solve some of the problems associated with OPA derivatization followed by LCEC; some very useful suggestions were given so that detection limits down to 300 pmol cm^{-3} could be achieved. These authors indicated that, with care, the method was applicable to concentrations below the detection limit of most fluorescent detectors.

The applications of OPA derivatization, coupled with LCEC, have been discussed in several short articles [134,135].

Isothiocyanate derivatives
Mahachi *et al.* [136] have described an alternative precolumn derivatization agent for the electrochemical detection of amino acids. From several isothiocyanates investigated, *p-N,N*-dimethylaminophenylisothiocyanate (DMAPI) was selected and this produced a substituted phenylthiohydantoin (PTH) when reacted with the amino acids; the substituted *p*-aminoaniline moiety on the PTH derivative was the electroactive species. This group undergoes oxidation at a glassy carbon electrode at +0.85 V versus Ag/AgCl; the mobile phase consisted of 0.1 M phosphate buffer–25% acetonitrile (pH 2 or 6). In this study, a mixture of 21 amino acids was separated with 80–98% recovery; calibration graphs were linear from 1 to greater than 150 ng injected and the detection limits were 0.5 to 1.0 ng.

Nitro-derivatives
It has been suggested that the determination of GABA is of particular importance because it acts as a neurotransmitter in the mammalian brain [137]. An LCEC method was described by Caudill *et al.* [137] which involved derivatization with TNBS. The authors showed that derivatization was quantitative even at low concentrators and derivatives could be electrochemically reduced with the consumption of 12 electrons at modest potentials (−0.8 V). Separations were performed with either a reversed-phase column or a strong cation exchange column; with the former, a chromatographic signal was clearly visible at about 16 min for only 2.3-pmol TNP-GABA. In this study, the method was successfully applied to rat brain and compared favourably with other methods.

3.2.3 Methods involving amperometric sensors and biosensors

Amino acids and glutathione
It is known that some enzymes are selective for a certain group of substrates, but are not specific within that group; for example, amino acid oxidases are specific for amino acids but not for particular amino acids [138]. It is also known that amino acid ocidases possess optical specificity for L-amino acids and for D-amino acids; L-amino acid oxidase (L-AAO) is specific for the former and D-amino acid oxidase (D-AAO) is specific for the latter. The general enzymatic reaction is:

$$\text{Amino acid} + O_2 + H_2O \xrightarrow{\text{AOO}} \text{2-oxo acid} + NH_3 + H_2O_2 \qquad (3.7)$$

Yao and Wasa [139] have exploited this reaction for the simultaneous determination of L- and D-amino acids using LCEC; the hydrogen peroxide produced in the reaction was used to monitor the amino acids. These authors employed working electrodes constructed by immobilizing either L-AAO or D-AAO onto the surface of platinum sheets by cross-linking the enzyme and BSA with glutaraldehyde; these electrodes were incorporated into separate flow-through cells. Fig. 3.29 shows the schematic configuration of the enzyme electrode in the amperometric detector.

Separation of the amino acids in mixtures was achieved on a reversed-phase column, after which the eluent from the column was adjusted to pH 8.0. The eluent stream was split in two and passed through the two cells; the D-isomers were detected in one stream and the L-isomers in the other. The detection limit was about 2 pmol for methionine, tyrosine (VII), leucine (IV) and phenylalanine (XI).

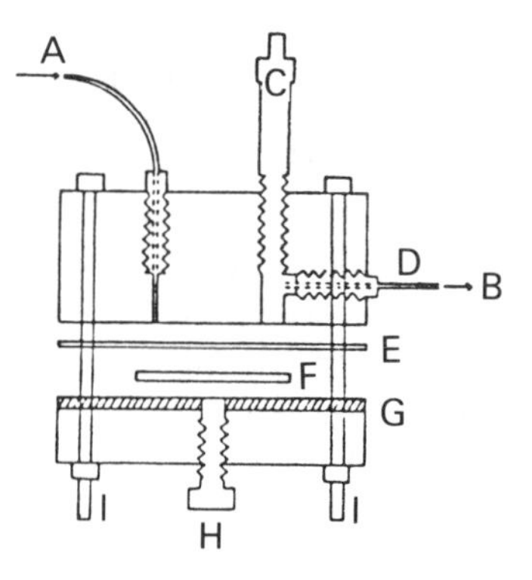

Fig. 3.29 — Schematic configuration of the enzyme electrode as an HPLC detector: (a) solution inlet, (b) solution outlet, (c) Ag/AgCl reference electrode, (d) auxiliary electrode (stainless steel tube), (e) spacer (0.05 mm thick giving a 2×16 mm channel), (f) enzyme-modified platinum sheet (20×20 mm, 0.1 mm thick), (g) silicone rubber, (h) stainless steel screw), (i) assembly screw. (Reproduced from [139] by permission of the copyright holders, Elsevier Science Publishers Physical Sciences and Engineering Division.).

An enzymatic method to measure L-glutamate was developed by Yamauci *et al.* [140]. L-Glutamate oxidase from streptomyces was immobilized on a cellulose acetate film in order to construct an enzyme oxygen electrode. A somewhat different approach to immobilizing LAAOs has been described by Ianniello and Yacynych [141]. These authors covalently bonded LAAO to a graphite electrode via a cyanuric chloride linkage; the resulting electrode was termed an immobilized enzyme chemically modified electrode (IECME). These sensors were able to detect the presence of L-phenylalanine through the reaction described above. (Eq. 3.7).

Recently, it has been shown that D- and L-amino acid oxidases can be oxidized directly at organic salt electrodes; this discovery has led to the indirect determination of several amino acids. Investigations carried out by Albery *et al.* [142] indicated that the mechanism of oxidation of the enzyme involved catalysis by the components of the organic conducting salt.

It has been reported [143] that glutathione sulphydryl oxidase catalyses the oxidation, by oxygen, of two molecules of reduced glutathione (GSH) to form one molecule each of oxidized glutathione (GSSG) and hydrogen peroxide; therefore, electrochemical detection of the latter can be used to measure GSH. This reaction was employed in a flow-injection method described by Satoh *et al.* [143] where the enzyme was immobilized onto oxirane-acrylic beads; these were contained in a small plastic column, which was placed upstream of the electrochemical detector. Detection of the hydrogen peroxide was carried out with a platinum electrode which was coated with a thin layer of cellulose acetate; the applied potential was set at +0.7 V versus Ag/AgCl. Calibration graphs were linear over the range 0.01 to 1 mM; L-cysteine gave a response which was 10% of that obtained for GSH.

3.3 CONCLUSIONS

The discussion in this chapter has been mainly concerned with the application of electroanalytical methods to the determination of a wide variety of amino acids in different samples.

Electroanalytical techniques have been applied to investigations into amino acid levels in biological fluids; in this particular area LCEC methods are showing great promise. One reason for this has been the development opf novel amperometric and coulometric detectors; working electrodes have been constructed from chemically modified carbon substrates, copper tubes, mercury, gold, gold amalgam and enzymes immobilized onto inert supports. Another reason for the growing interest in LCEC has been the discovery that derivatization methods can easily be used to convert a variety of electroinactive amino acids into electroactive derivatives; this has greatly increased the potential of LCEC in biomedical analysis.

In the area of biomedical analysis, LCEC methods have been used to measure simultaneously circulating levels of amino acids and their metabolites; such determinations are of obvious importance in studies of metabolic disorders. Clearly, these methods can also be readily applied to studies where therapeutic levels may be encountered. It seems likely that in the future, methods involving this technique will become increasingly important, especially where high sensitivity is required.

A variety of examples have been included to show the versatility of voltammetric procedures. Stripping voltammetry is the most sensitive of these techniques; methods utilizing copper amalgam electrodes have shown promise for low levels of drugs related to cysteine (e.g. penicillins); detection limits as low as 2×10^{-10} M have been obtained. Derivatization, followed by differential polarography, has been successfully used in studies on protein degradation. For complex amino acid mixtures it has been possible to improve the selectivity of polarography by the incorporation of a thin-layer chromatographic procedure. Amino acids containing oxidizable groups may be determined voltammetrically using carbon electrodes. It has also been shown that a chemical modifying agent reduces the overpotential for the oxidation of some sulphur-containing amino acids and related drugs; it may be possible to apply such techniques to other similar species.

Peptides which contain electroactive amino acids are clearly suitable candidates for analysis by electrochemical methods. Perhaps the most often studied and therefore repeatedly described here has been glutathione. DPP was capable of monitoring whole blood levels of this peptide. More recently, electrochemical sensors containing a chemically modifying agent have shown promise. However, for circulating plasma levels an LCEC procedure would be particularly appropriate. In addition to glutathione, low levels of some neurotransmitter peptides containing tyrosine and/or tryptophan residues have been successfully monitored by LCEC methods; such methods have been used for investigations on rat brain. Obviously, these methods for peptide analysis could find application in other areas of biomedical research.

The determination of proteins continues to be an area of considerable interest. In Eastern Europe polarographic methods that exploit the catalytic reactions of proteins have been developed; these have been successfully applied to a number of clinical investigations including investigations of mental disorders, and cirrhosis. It was recently shown that a catalytic reaction could be used as the basis for a rapid automated continuous-flow system where up to 120 samples of serum an hour could be analysed for proteins. Recently, however, other electrochemical methods have been developed to determine proteins. AdSV has been readily applied to the determination of bovine serum albumin (BSA), human serum albumin (HSA), and anti-HSA as well as other proteins. It has also been employed to follow protein/protein interactions as well as protien/monosaccharide interactions. It is possible that biosensors for some of these molecules may result from these basic studies.

REFERENCES

[1] C. Barrett (Ed.), *Chemistry and Biochemistry of the Amino Acids*, Chapman and Hall, London, 1985.
[2] G. L. Blackburn, J. P. Grant and V. R. Young (Eds), *Amino Acids: Metabolism and Medical Applications*, John Wright, PSG Inc., Boston, 1983.
[3] W. K. Jacoby and O. W. Griffiths (Eds) Sulfur and Sulfur Amino Acids, in *Methods in Enzymology*, vol. 143, S. P. Colowick and N. O. Kaplan (Editors in Chief), Academic Press, 1987.
[4] S. Kaufman (Ed.), Metabolism of Aromatic Amino Acids and Amines, in, *Methods in Enzymology*, S. P. Colowick and N. O. Kaplan (Editors in Chief), Academic Press, 1987.
[5] C. H. Letendre, K. Nagaiah and G. Guroff, in *Biochemistry of Brain*, S. Kumar (Ed.), p. 343, Pergamon Press, 1980.
[6] J. Marka, *The Treatment of Parkinsonism with L-Dopa*, Elsevier, New York, 1974.
[7] D. A. Bender, *Amino Acid Metabolism*, John Wiley and Sons, Chichester, 1975.
[8] J. Mitchell and C. J. Coscia, *J. Chromatogr.*, 1978, **145**, 295.
[9] G. W. Scheiffer, *J. Pharm. Sci.*, 1979, **68**, 1299.
[10] J. G. ALlen, *Adv. Neurol.*, 1973, **3**, 131.
[11] W. Lee, *Biochem. J.*, 1971, **121**, 563.
[12] C. A. Mairesse-Ducarmoise, G. J. Patriarche and J. L. Vandenbalck, *Anal. Chim. Acta*, 1974, **71**, 165.
[13] C. A. Mairesse-Ducarmoise, G. J. Patriarche and J. L. Bandenbalck, *J. Pharm. Belg.*, 1976, **31**, 169.
[14] A. D. Woolfson, J. S. Millership and E. I. Karim, in *Electrochemistry Sensors and Analysis*, M. R. Smyth and J. G. Vos (Eds), p. 379, Elsevier, Amsterdam, 1986.
[15] W. Stricks and I. M. Kolthoff, *J. Am. Chem. Soc.*, 1952, **74**, 4646.
[16] C. A. Mairesse-Ducarmoise, G. J. Patriarche and J. L. Vandenbalck, *Anal. Chim. Acta*, 1975, **76**, 299.
[17] C. A. Mairesse-Ducarmoise, PhD Thesis, Universite Libre De Bruxelles, 1977.
[18] G. J. Patriarche and J.-C. Vire, in *Electroanalysis in Hygiene, Environmental, Clinical and Pharmaceutical Chemistry*, W. F. Smyth (Ed.), Elsevier, Amsterdam, 1980, pp. 209–225.
[19] J. P. Hart, in *Investigative Microtechniques in Medicine and Biology*, J. Chayen and L. Bitensky (Eds), Marcel Dekker, 1984, pp. 199–250.
[20] R. Cecil and P. D. J. Weitzman, *Biochem. J.*, 1964, **93**, 1.
[21] P. D. J. Weitzman, *Biochim. Biophys. Acta*, 1965, **107**, 146.
[22] M. T. Stankovich and A. J. Bard, *J. Electroanal. Chem.*, 1977, **85**, 173.
[23] T. Stankovich and A. J. Bard, *J. Electroanal. Chem.*, 1978, **86**, 189.
[24] E. Palecek, in *Topics in Bioelectrochemistry and Bioenergetics*, G. Milazzo (Ed.), Vol. 5, John Wiley and Sons, 1983, pp. 65–157.
[25] W. Hertl, *Anal. Biochem.*, 1987, **164**, 1.
[26] A. Sanchez Perez, F. Lucena Conde and J. Hernandez Mendez, *J. Electroanal. Chem.*, 1976, **74**, 339.
[27] J. Hernandez Mendez, A. Sanchez Perez and F. Lucena Conde, *J. Electroanal. Chem.*, 1975, **6**, 53.
[28] R. Brdicka, *Collect Czech. Chem. Commun.*, 1933, **5**, 112.
[29] R. Brdicka, *Collect Czech. Chem. Commun.*, 1933, **5**, 148.
[30] E. Palecek, V. Brabec, F. Jelen and Z. Pechan, *J. Electroanal. Chem.*, 1977, **75**, 471.
[31] E. Palecek and Z. Pechan, *Anal. Biochem.*, 1971, **42**, 59.
[32] V. Ueno, F. Kuraishi and S. Vematsu, *T. Tsuruoka, Folia, Psychiatr. Neurol*, Japan, 1968, **22**, 167.
[33] L. E. Moysenko, *Neuropatol, Psikhiatr. IM. SS. Korsakova*, 1972, **72**, 854.
[34] L. E. Moysenko and S. N. Vlasenko, *Vrach Delo.*, 1973, **1**, 139.
[35] I. D. Mansurova, *Lab. Delo.*, 1974, **7**, 714.

[36] B. Z. Chowdry, in *Polarography of Molecules of Biological Significance*, W. F. Smyth (Ed.), Academic Press, London, 1979, pp. 169–201.
[37] G. Ruttkay-Nedecky and B. Bezuch, *J. Mol. Biol.*, 1971, **55**, 101.
[38] G. Ruttkay-Nedecky and A. Anderleova, *Nature*, 1967, **213**, 564.
[39] G. Ruttkay-Nedecky , B. Bezuchand and V. Vesela, *Bioelectrochem. Bioenerget.*, 1977, **4**, 399.
[40] G. Ruttkay-Nedecky and V. Vesela, *Acta Virol. Engl. Ed.*, 1977, **21**, 365.
[41] P. W. Alexander and M. H. Shah, *Talanta*, 1979, **26**, 97.
[42] P. W. Alexander, R. Hoh and L. E. Smythe, *J. Electroanal. Chem.*, 1977, **80**, 143.
[43] K. Kano, T. Konse and T. Kubota, *Anal. Sci.*, 1986, **2**, 507.
[44] M. Brezina and P. Zuman, *Polarography in Medicine, Biochemistry and Pharmacy*, Interscience Publishers, New York, 1958.
[45] R. Brdicka, M. Brezina and V. Kalous, *Talanta*, 1965, **12**, 1149.
[46] J. Homolka, in *Methods in Biochemical Analysis*, D. Glick (Ed.), J. Wiley, New York, Vol. 19, 1971, p. 435.
[47] H. Berg, in *Topics in Bioeelctrochemistry and Bioenergetics*, G. Milazzo (Ed.), Vol. 1, Wiley-Interscience, London, 1983.
[48] M. A. Al-Hajjaji, *Anal. Chim. Acta*, 1984, **157**, 31.
[49] T. Okuyama and K. Satake, *J. Biochem.*, 1960, **47**, 454.
[50] S. L. Snyder and P. Z. Sobocinski, *Anal. Biochem.*, 1975, **64**, 284.
[51] M. A. Hajjaji, *Anal. Chim. Acta*, 1986, **181**, 227.
[52] G. Ramin Ramos, A. R. Mauri Aucejo, M. C. Garcia Alvarez-coque and C. Mongay Gernandez, *Microchem. J.*, 1987, **36**, 113.
[53] T. M. Florence, *J. Electroanal. Chem.*, 1979, **97**, 219.
[54] I. E. Davidson, in *Polarography of Molecules of Biological Significance*, W. F. Smyth (Ed.), Academic Press, London, 1979, pp. 127–165.
[55] W. F. Smyth, in *Electroanalysis in Hygiene, Environmental, Clinical and Pharmaceutical Chemistry*, W. F. Smyth (Ed.), Elsevier, Amsterdam, 1980, pp. 271–286.
[56] J. T. Stock and R. E. Larson, *Anal. Chim. Acta*, 1982, **138**, 371.
[57] H. Berge and P. Jeroschewski, *Z. Anal. Chem.*, 1965, **212**, 278.
[58] F. Vydra, K. Stulik and E. Julakova, *Electrochemical Stripping Analysis*, Ellis Horwood, Chichester, 1976.
[59] M. Donten and Z. Kublik, *Anal. Chim. Acta*, 1986, **185**, 209.
[60] V. Forsman, *Anal. Chim. Acta*, 1983, **146**, 71.
[61] J. Wang, V. Villa and T. Tapia, *Bioelectrochem. Bioengerg.*, 1988, **19**, 39.
[62] J. R. Flores, R. O'Kennedy and M. R. Smyth, *Anal. Chim. Acta*, 1988, **212**, 355.
[63] J. R. Flores, R. O'Kennedy and M. R. Smyth, *Analyst*, 1988, **113**, 525.
[64] J. R. Flores and M. R. Smyth, *J. Electroanal. Chem.*, 1987, **235**, 317.
[65] M. R. Smyth, E. Buckley, J. R. Flores and R. O'Kennedy, *Analyst*, 1988, **113**, 31.
[66] J. R. Flores, R. O'Kennedy and M. R. Smyth, *Biosensors*, 1988, **4**, 1.
[67] C. R. Linders, B. J. Vincke, J.-C. Vire, J.-M. Kauffmann and G. J. Patriarche, *J. Pharm. Belg.*, 1985, **40**, 27.
[68] N. Verbiese-gerard, J. M. Kauffmann, M. Hanocq and L. Molle, *J. Electroanal. Chem.*, 1984, **170**, 243.
[69] N. V. Shvedene, N. M. Sheina, L. I. Tkacheva and L. Yu. Parinova, *Vestn. Mosk. Univ. Ser. 2: Khim.* 1988, **29**, 77.
[70] M. K. Halbert and R. P. Baldwin, *Anal. Chem.*, 1985, **57**, 591.
[71] M. K. Halbert and R. P. Baldwin, *J. Chromatogr.* 1985, **345**, 43.
[72] S. A. Wring, J. P. Hart and B. J. Birch, *Analyst*, 1989, **114,** 1563.
[73] M. D. Hawley, S. V. Tatawadi, S. Piekarski and R. N. Adams, *J. Am. Chem. Soc.*, 1967, **89**, 447.
[74] R. C. Causon, M. J. Brown, K. L. Leenders and L. Wolfson, *J. Chromatogr.*, 1983, **277**, 115.
[75] R. M. Riggin, R. L. Alcorn and P. T. Kissinger, *Clin. Chem.*, 1976, **22**, 782.
[76] E. Nissinen and J. Taskinen, *J. Chromatogr.*, 1982, **231**, 459.
[77] M. F. Beers, M. Stern, H. Hurtig, G. Melvin and A. Scarpa, *J. Chromatogr.*, 1984, **336**,380.
[78] Y. Michotte, M. Moors, D. Deleu, P. Herrgobts and G. Ebinger, *J. Pharm. Biomed. Anal.*, 1987, **5**, 659.
[79] A. Baruzzi, M. Contin, F. Albini and R. Riva, *J. Chromatogr.*, 1986, **375**, 165.
[80] F. Boomsma, F. A. J. van der Hoorn, A. J. Man In't Veld and M. A. D. H Schalekamp, *Clin. Chim. Acta*, 1988, **178**, 59.
[81] T. Ishimitsu and S. Hirose, *J. Chromatogr.*, 1985, **337**, 239.
[82] C. R. Benedict and M. Risk, *J. Chromatogr.*, 1984, **317**, 27.
[83] A. Shum, G. R. Van Loonand and M. J. Sole, *Life Sci.*, 1982, **31**, 1541.
[84] T. A. Last, *Anal. Chim. Acta*, 1983, **155**, 287.

[85] L. D. Hutchins, J. Wang and P. Tuzhi, *Anal. Chem.*, 1986, **58**, 1019.
[86] J. Wang and L. D. Hutchins, *Anal. Chem.*, 1985, **57**, 1536.
[87] J. W. Dieker, W. E. Van der Linden and H. Poppe, *Talanta*, 1979, **26**, 511.
[88] K. Stuluk and V. Pacakova, *Electroanalytical Measurements in Flowing Liquids*, Ellis Horwood, Chichester, 1987.
[89] V. Pacakova and K. Stulik, *Die Nahrung*, 1985, **29**, 651.
[90] D. D. Koch and P. T. Kissinger, *J., Chromatogr.*, 1979, **164**, 441.
[91] G. A. Qureshi and S. Gokmen, *Chim. Acta Turc.*, 1986, **14**, 299.
[92] R. J. Martin, B. A. Bailey and R. G. H. Downer, *J. Chromatogr.*, 1983, **278**, 265.
[93] A. Sauter and W. Frick, *Anal. Biochem.*, 1983, **133**, 307.
[94] R. Dawson, J. P. Steves, J. F. Lorden and S. Oparil, *Peptides*, 1985, **6**, 1173.
[95] J. Di Bussolo, *Internatl. Biotech. Lab.*, 1984, **2**, 14.
[96] G. Carro-Ciampi, P. G. Hunt, C. J. Turner and P. J. Wells, *J. Pharmacol. Methods*, 1988, **19**, 75.
[97] M. K. Halbert and R. P. Baldwin, *J. Chromatogr., Biomed. Appl.*, 1985, **345**, 43.
[98] S. A. Wring, J. P. Hart and B. J. Birch, *Analyst*, 1989, **114**, 1571.
[99] J. Wang, T. Golden, K. Varughese and I. El-Rayes, *Anal. Chem.*, 1989, **61**, 508.
[100] C. R. Loscombe, G. B. Cox and J. A. W. Dalziel, *J. Chromatogr.*, 1978, **166**, 403.
[101] P. W. Alexander, P. R. Haddad, G. K. C. Low and C. Maitra, *J. Chromatogr.*, 1981, **209**, 29.
[102] P. W. Alexander and C. Maitra, *Anal. Chem.*, 1981, **53**, 1590.
[103] W. Th. Kok, H. B. Hanekamp, P. Bos and R. W. Frei, *Anal. Chim. Acta*, 1982, **142**, 31.
[104] W. Th. Kok, U. A. Th. Brinkman and R. W. Frei, *J. Chromatoigr.*, 1983, **256**, 17.
[105] W. Th. Kok, U. A. Th. Brinkman and R. W. Frei, *J. Pharm. Biomed. Anal.*, 1983, **1**, 369.
[106] K. Stulik, V. Pacakova, M. Weingart and M. Posolak, *J. Chromatogr.*, 1986, **367**, 311.
[107] K. Stulik, V. Pacakova and M. Podolak, *J. Chromatog.*, 1984, **298**, 225.
[108] K. Stulik, V. Pacakova and G. Jokuszies, *J. Chromatog.*, 1988, **436**, 334.
[109] A. Ichimura, M. Nakatsuka, K. Ogura and T. Kitagawa, *Nippon Kagaku Kaishi*, 1986, 987, 992.
[110] D. L. Rabenstein and R. Saetre, *Anal. Chem.*, 1977, **49**, 1036.
[111] R. Saetre and D. L. Rabenstein, *Anal. Biochem.*, 1978, **90**, 684.
[112] R. Saetre and D. L. Rabenstein, *Anal. Chem.*, 1978, **50**, 276.
[113] R. Eggli and R. Asper, *Anal. Chim. Acta,* 1978, **101**, 253.
[114] L. A. Allison, *Current Separations*, 1982, **4**, 38.
[115] L. A. Allison and R. Shoup, *Anal. Chem.*, 1983, **55**, 8.
[116] J. P. Richie and C. A. Lang, *Anal. Biochem.*, 1987, **163**, 9.
[117] J. M. Walshe, *Br. J. Hosp. Med.*, 1970, **4**, 91.
[118] J. C. Crawhall, D. Lecavalier and P. Ryan, *Biopharm. Drug Dispos.*, 1979, **1**, 73.
[119] M. Aylward, J. Maddock and P. M. Dewland, *Eur. J. Respir. Dis.*, 1980, Supl. III, 81.
[120] O. H. Drummer, N. Christophidis, J. D. Horowitz and W. J. Louis, *J. Chromatogr.*, 1986, **374**, 251.
[121] E. G. Demaster, F. N. Shiroto, B. Redfern, D. J. W. Goon and H. T. Nagasawa, *J. Chromatogr.*, 1984, **308**, 83.
[122] R. F. Bergstrom, D. R. Kay and J. G. Wagner, *J. Chromatogre.*, 1981, **222**, 445.
[123] F. Kreuzig and J. Frank, J. Chromatogr., 1981, **218**, 615.
[124] M. H. Joseph and P. Davies, *J. Chromatogr.*, 1983, **277**, 125.
[125] K. Bratin, C. L. Blank, I. S. Krull, C. E. Lunte and R. E. Shoup, *Internatl. Lab.*, 1984, **14**, 24.
[126] P. Lindroth and K. Mopper, *Anal. Chem.*, 1979, **51**, 1667.
[127] L. A. Allison, G. S. Mayer and R. E. Shoup, *Anal. Chem.*, 1984, **56**, 1089.
[128] H. R. Zielke, *J. Chromatogr.*, 1985, **347**, 320.
[129] A. H. Hikal, G. W. Lipe, W. Slikker, A. C. Scallet, S. F. Ali and G. D. Newport, *Life Sci.*, 1988, **42**, 2029.
[130] S. J. Ziegler and D. Sticher, HRC CC, *J. High Res. Chromatogr., Chromatogr. Commun.*, 1988, **11**, 639.
[131] E. G. Demaster, F. N. Shirota, B. Redfern, D. J. W. Goon and H. T. Nagasawa, *J. Chromatogr.*, 1984, **308**, 83.
[132] W. Buchburger and K. Winsauer, *Anal. Chim. Acta*, 1987, **196**, 251.
[133] J. A. Hoskins, S. B. Holiday and F. F. Holiday, *J. Chromatogr.*, 1986, **375**, 129.
[134] W. Jacobs, *Current Sep.*, 1986, **7**, 39.
[135] Z. K. Shihabi, *J. Liq. Chromatogr.*, 1985, **8**, 2805.
[136] T. J. Mahachi, R. M. Carlson and D. P. Poe, *J. Chromatogr.*, 1984, **298**, 279.
[137] W. L. Caudill, G. P. Houck and R. M. Wightman, *J. Chromatogr.*, 1982, **227**, 331.
[138] G. G. Guilbault and G. J. Lubrano, *Anal. Chim. Acta*, 1974, **69**, 183.
[139] T. Yao and T. Wasa, *Anal. Chim. Acta*, 1988, **209**, 259.
[140] H. Yamauchi, H. Kusakabe and Y. Midorikawa, *Nippon Shoyu Kenkyusho Zasshi*, 1987, **13**, 8.
[141] R. M. Ianniello and A. M. Yacynych, *Anal. Chem.*, 1981, **53**, 2090.
[142] W. J. Albery, P. N. Bartlett and A. E. G. Cass, *Phil. Trans. R. Soc. Lond.*, 1987, B 316, 107.
[143] I. Satoh, S. Arakawa and A. Okamoto, *Anal. Chim. Acta*, 1988, **214**, 415.

4

Vitamins

4.1 INTRODUCTION

There is a growing need for sensitive, selective and reliable methods for the analysis of vitamins in such areas as nutrition and pharmaceutical chemistry, as well as in clinical chemistry and research [1,2]. However, these determinations may present difficult analytical problems for a variety of reasons; for example, in blood, interferences between structurally similar vitamers can occur; also, interference from the presence of other naturally occurring compounds may exist. In addition, the low endogenous levels present put considerable constraints on both the sensitivity and selectivity of potential analytical procedures [3,4].

As indicated in previous sections, modern electroanalytical techniques can be used to overcome many of the problems associated with difficult analyses in complex matrices. Therefore, since most of the vitamins are electroactive it is perhaps not surprising that many of these techniques have been applied to their determination. In this chapter examples have been selected to illustrate the versatility of these electroanalytical methods in vitamin analysis.

The vitamins can be broadly divided into fat-soluble and water-soluble groups and the following discussion has been divided accordingly.

4.2 ELECTROANALYSIS OF FAT-SOLUBLE VITAMINS

4.2.1 Vitamin A

4.2.1.1 Methods involving polarographic and voltammetric techniques

Vitamin A consists of a large group of structurally similar compounds. Within this group are a subgroup known as vitamin A_1 and these have the structures shown in

I–VI: where R=CH_2OH in *trans*-retinol (I) and in 13-*cis*-retinol (II); R=CHO in retinal (III); R=COOH in retinoic acid (IV); R=CH_3COO in retinyl acetate (V); R=$CH_3(CH_2)_{14}COO$ in retinyl palmitate (VI); all-*trans* retinol has the highest vitamin A activity. The compounds belonging to this group are essential for human growth and are important for good vision. Recent reports suggest that this vitamin might have additional roles such as in protecting against cancer and in the treatment of skin disorders [5].

Older electrochemical methods for the analysis of vitamin A compounds were based on reduction reactions occurring at a DME. Since this vitamin group is fat-soluble, high concentrations of organic solvents were required for dissolution; quaternary ammonium salts have been employed as supporting electrolytes.

Takahashi and Tachi [6] investigated the electrochemical reduction behaviour of all-*trans* retinol (I) and retinyl palmitate (VI) in a medium consisting of 0.1-M tetrabutylammonium iodide in 60% benzene–acetonitrile. These authors indicated that the two vitamers were initially reduced in a diffusion-controlled two-electron step to produce a species with four conjugated double bonds; the half-wave potentials were -1.41 V and -1.23 V versus a mercury pool for (I) and (VI) respectively. Further reduction occurred by the addition of two electrons across a conjugated double bond to produce a second polarographic wave with an $E_{1/2}$ value of -1.65 V for both compounds. A third wave appeared on the d.c. polarograms, which was again the result of a similar two-electron reduction process; in this case the waves were about 200 mV more cathodic and were only ill-defined. Similar results for compound (I) were obtained by Kuta [7]; however, retinal (III) did not appear to follow the same mechanism of reduction as the other vitamers. The author indicated that this compound may initially be reduced with the consumption of one electron, which could then undergo a dimerization reaction.

Detailed studies on the electrochemical reduction of (I) and (III) have been carried out at a platinum working electrode in THF-containing tetrabutylammonium chlorate [8]; in this study it was found that retinal was reduced at various double bonds, resulting in many products including retinol.

Calibration studies on (I) and (VI) showed that the limiting diffusion current was linearly dependent upon concentration over the range 5×10^{-5} to 10^{-3} M [6]. However, it should be mentioned that modern voltammetric techniques such as DPP and square wave voltammetry [9,10] would probably extend the detection limit to about 10^{-7}–10^{-8} M.

The electroreduction of vitamin A has been exploited for the determination of the vitamin in an oil-based pharmaceutical product containing a ten-fold excess of vitamin D_2 [11]. In this method the oscillographic determination was performed with a supporting electrolyte comprising 1% tetraethylammonium bromide in isopropyl alcohol–dimethylformamide (1:1).

In addition to reduction processes some vitamers of the A group have been found to undergo oxidation reactions at carbon electrodes. The present author and coworkers [12] performed detailed investigations into the voltammetric behaviour of all-*trans* retinol at a glassy carbon electrode; one of the reasons for this study was to obtain the optimum conditions for a method involving LCEC (discussed later). In this investigation cyclic voltammetry was performed on the vitamin in solutions containing 95% methanolic acetate buffers of different pH values. In all instances the

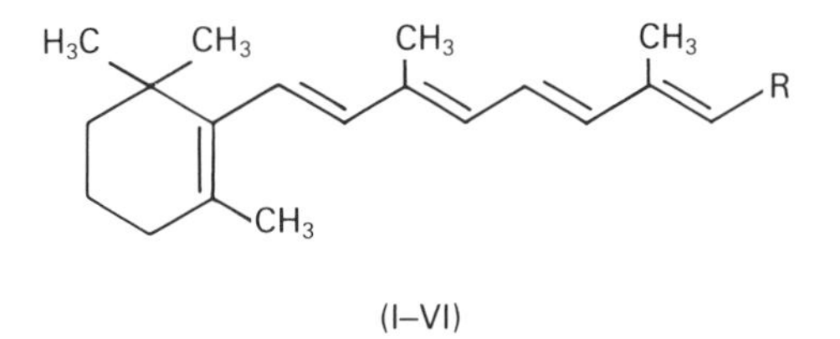

(I–VI)

first anodic scan gave two peaks, 1 and 2 (Fig. 4.1), but no cathodic peaks appeared on the reverse scans; this suggested that the overall oxidation process was irreversible. For peak 1 the variation of peak current and peak potential with the apparent pH of the supporting electrolyte is shown in Fig. 4.2. As may be seen, a break appears in the E_P versus pH plot indicating a pK_a value of 6.7. Below this break the E_P value obeys the relationship

$$E_P(\text{V}) = +0.976 - 0.0335\ \text{pH} \tag{4.1}$$

The results obtained in this study indicated that the overall number of electrons transferred in the reaction corresponding to peak 1 was likely to be three; peak 2 was probably due to a one-electron oxidation process. This is in agreement with the findings of MacCrehan and Schonberger [13] for peak 1; although there is a difference in the magnitude of peak 2, this may be the result of the different buffer and solvent system employed. It was suggested by MacCrehan and Schonberger [13] that the oxidation mechnism involved loss of electrons from the unsaturated hydrocarbon chain; this appeared to be followed by a variety of chemical reactions for which several different products have been proposed.

Investigations into the oxidative behaviour of some A vitamers has also been carried out using a carbon paste electrode [14]. In this study, the supporting electrolyte consisted of 75% ethanol–0.01-M sulphuric acid and compounds (I), (III), (IV), (V) and (VI) exhibited anodic peaks at +0.67, +0.80, +0.78 +0.71 and +0.85 V respectively; the authors did not mention whether a second volatammetric peak appeared. These voltammetric peaks could be utilized for the determination of the vitamers in several different pharmaceutical products; either direct solvent extraction, or a simple sample treatment followed by solvent extraction could be used to remove potential interferences.

4.2.1.2 Methods involving LCEC

A recent study was carried out by MacCrehan and Schonberger [13] to investigate the possibility of separating six isomers of retinol, including the all-*trans* isomer (I), by reversed-phase HPLC. The best resolution was obtained with a mobile phase containing butanol–methanol–water with ammonium acetate buffer (pH 3.2). These workers used both UV and electrochemical detectors which appeared to give similar results.

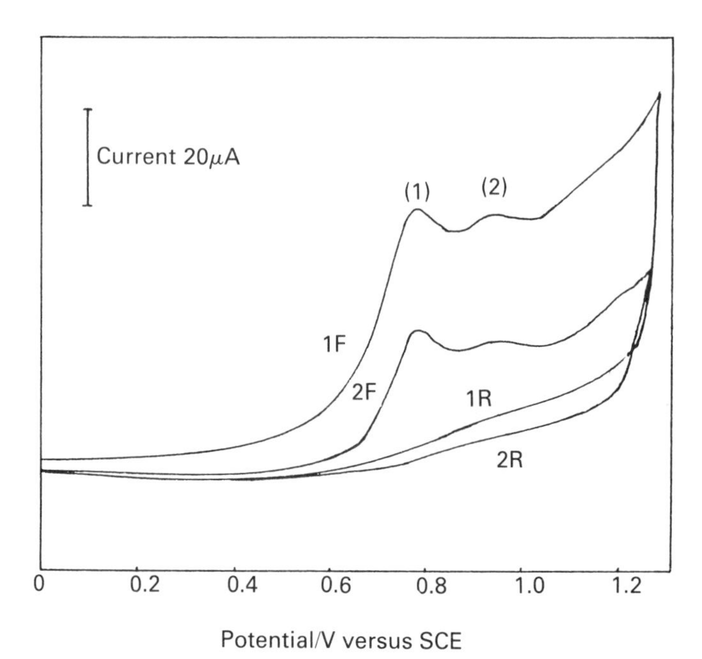

Fig. 4.1 — Cyclic voltammogram of 5×10^{-4}-M all-*trans*-retinol in 95% methanol–0.05-M acetate buffer solution (pH 5.0) using a glassy carbon electrode. Initial potential, 0 V; scan rate, 50 mV s^{-1}. 1F and 2F are the first and second forward scans respectively; 1R and 2R are the first and second reverse scans respectively. (Reproduced from [12] by permission of the copyright holders, Royal Society of Chemistry.)

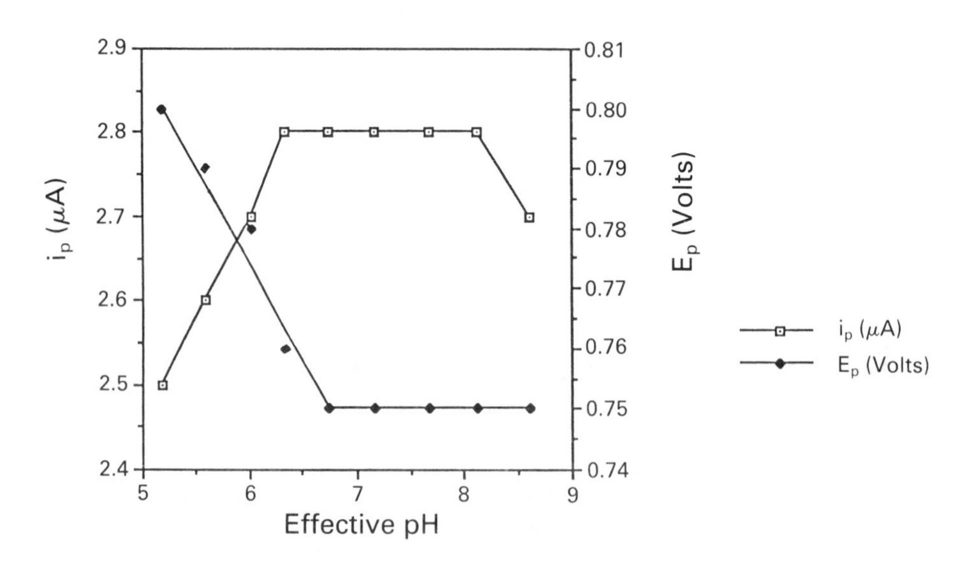

Fig. 4.2 — Effect of pH on peak current and peak potential for peak 1 using 5.3×10^{-5}-M all-*trans*-retinol in 95% methanol–0.05-M acetate buffer (pH 5.0). (Reproduced from [12] by permission of the copyright holders. Royal Society of Chemistry.)

The present author and coworkers [12] have also reported on the development of an LCEC method for (I) which was required to be suitable for the measurement of both normal and, in particular, depressed circulating levels in human serum. It should be added that all-*trans* retinol is considered to be the predominant form of the vitamin in the circulation and for this reason it was the only isomer studied. From the initial voltammetric studies mentioned earlier [12], the optimum supporting electrolyte consisted of 95% methanol–0.075-M pH 5.0 acetate buffer. Hence, this was used as the mobile phase in the LCEC method which we developed. In this study, a wall-jet cell was used which contained a glassy carbon working electrode; the optimum applied potential was found to be +1.1 V. Fig. 4.3 shows the LCEC chromatograms for a serum sample spiked with (I) and for the control serum itself. The endogenous concentration of all-*trans* retinol in this sample was 676 ng cm^{-3}; the recovery was calculated to be 101%. The method was next applied to a group of normal subjects and a group of patients suffering with cirrhosis of the liver; the mean circulating level of the former was found to be 502 ng cm^{-3} and the later 275 ng cm^{-3}. These values were significantly different. The results were compared to retinol binding protein (RBP) levels for the cirrhotics and the correlation was significant [15]; since the RBP method is used as a marker for the disease, and is rather time-consuming, it might be worthwhile considering the LCEC as an alternative method for monitoring possible cases.

In a very recent study Hart and Jordan [16] investigated the possibility of using amperometric detection to monitor retinol in pharmaceutical products. Obviously, the LCEC method described earlier should also be applicable to this type of analysis. However, it did occur to us that it might be feasible to apply a less sophisticated method using FIA with a short column containing a reversed-phase packing material; we considered this to be low-pressure LCEC. In order to obtain a suitable retention time it was necessary to reduce the amount of methanol in the mobile phase to 65%; however, the buffer composition was the same as that described above. Fig. 4.4 shows the chromatogram obtained after sample saponification, followed by dilution with mobile phase and direct injection. Clearly the method was suitable for this particular pharmaceutical preparation and is likely to be adaptable to other similar preparations.

4.2.2 Vitamin D

Vitamin D_2 (ergocalciferol, VII) and vitamin D_3 (cholecalciferol, VIII) are the two major forms of vitamin D and have equal biological potency; they may be used in the prevention of the bone disease known as rickets [17].

The two D vitamers have been found to undergo reduction at a DME; the mechanism was considered to involve the addition of two electrons and two protons to one of the double bonds [18]. The supporting electrolytes used in polarographic studies of vitamin A have also been employed for these two substances. In 0.1-M tetrabutylammonium hydroxide compounds VII and VIII gave a single cathodic wave with $E_{1/2}$ values of −2.01 and −2.25 respectively [18]. Menicagli *et al.* [19] showed that it was possible to utilize the cathodic wave of D_2 for its determination in oils without interference from vitamin A.

In addition to reduction processes at mercury electrodes, vitamin D has been found to undergo oxidation reactions at carbon electrodes. Atuma [20,21] utilized

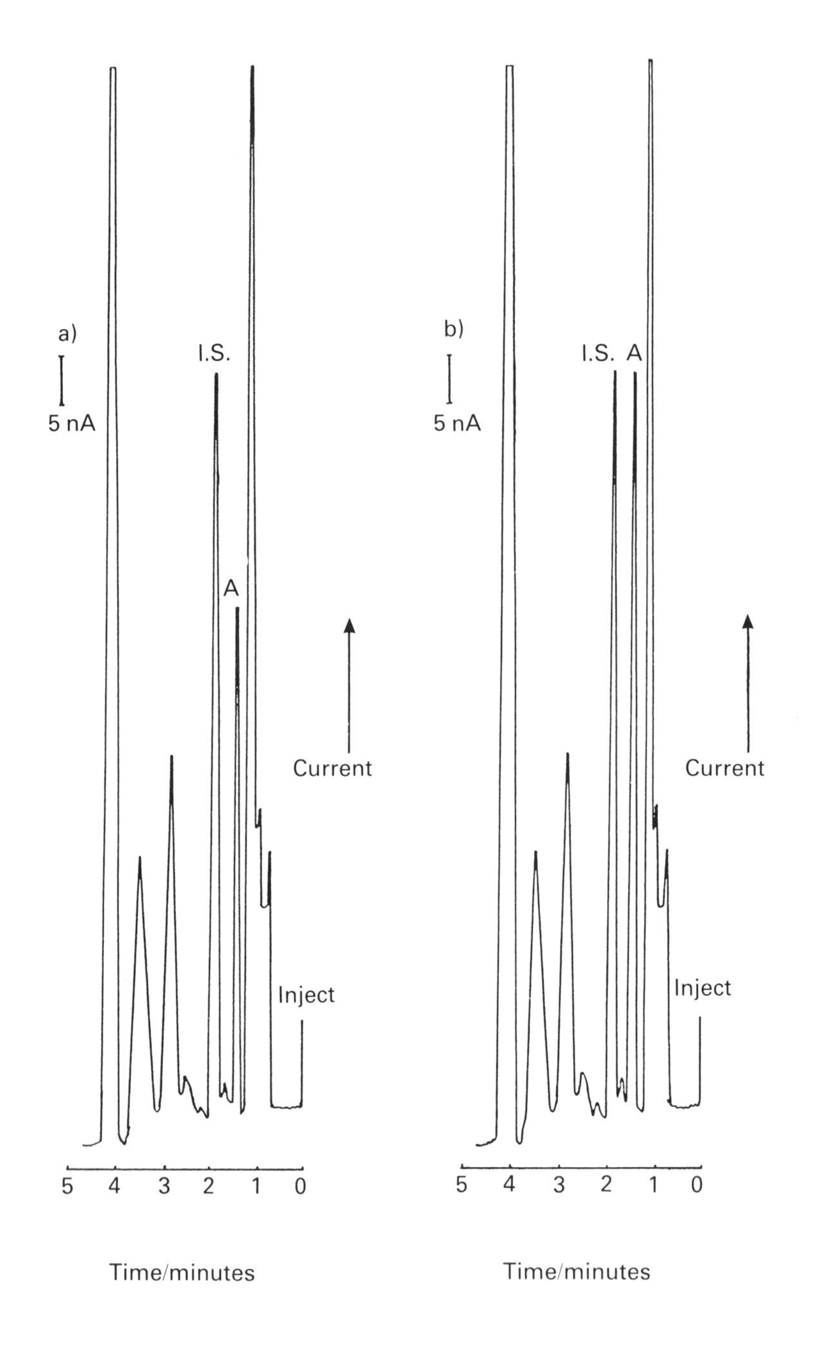

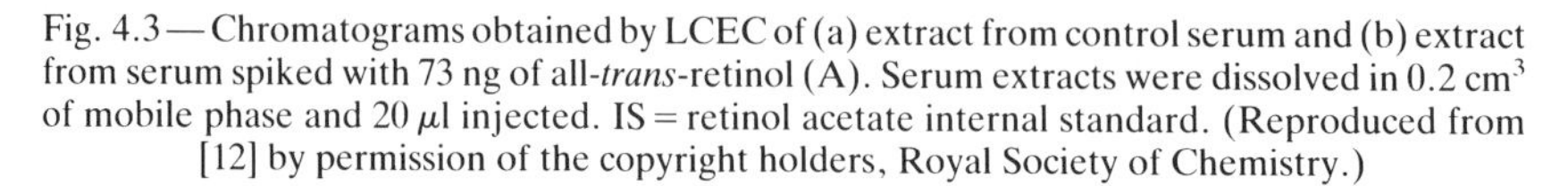

Fig. 4.3 — Chromatograms obtained by LCEC of (a) extract from control serum and (b) extract from serum spiked with 73 ng of all-*trans*-retinol (A). Serum extracts were dissolved in 0.2 cm^3 of mobile phase and 20 μl injected. IS = retinol acetate internal standard. (Reproduced from [12] by permission of the copyright holders, Royal Society of Chemistry.)

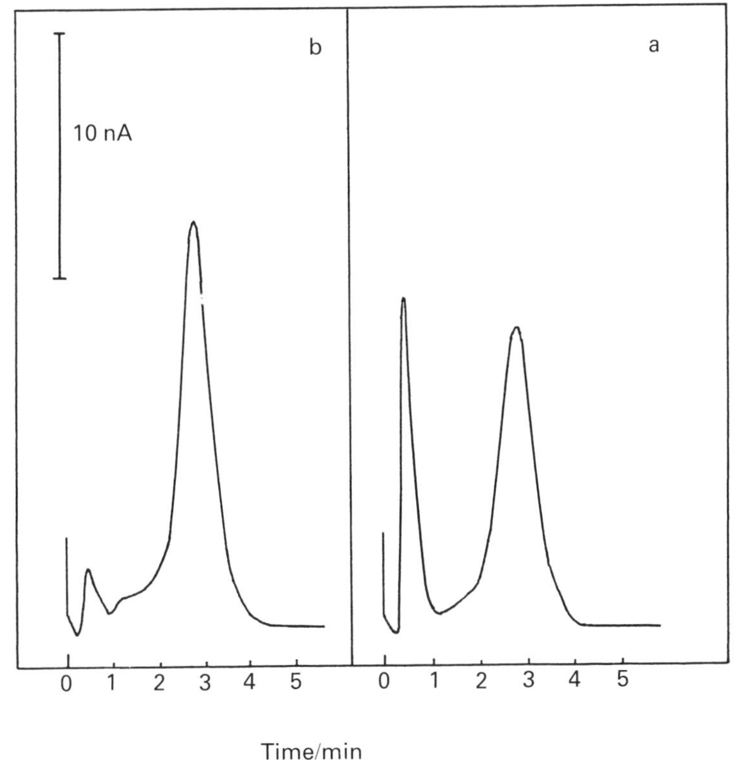

Fig. 4.4 — Low-pressure LCEC chromatograms of (a) Abidec drops following hydrolysis and dilution with mobile phase, (b) 60 ng retinol standard dissolved in mobile phase. Current range 50 nA. (Reproduced from [16] by permission of the copyright holders, Royal Society of Chemistry.)

this reaction and employed a glassy carbon electrode to monitor the vitamin D present in admixture with vitamin A.

Very recently, a method was reported for the voltammetric analysis of vitamin D_3 using a rotating disc glassy carbon electrode [22]. The supporting electrolyte consisted of methanol containing 0.075-M $LiClO_4$, and the resulting oxidation process was found to be irreversible. The differential-pulse, or d.c., waveform was applied to the working electrode and the anodic peak occurred at about +1.1 V (versus SCE) (Fig. 4.5). The calibration graph was linear in the range 2×10^{-6} M to 2×10^{-4} M; this could be used to calculate the content of pharmaceuticals, or alternatively the method of standard addition could be employed.

Another study has been carried out by the same research group on a metabolite of vitamin D_3, 25-hydroxy-vitamin D_3 (25-OH-D_3) [23]. In this study a rotating disc glassy carbon electrode was again used to determine 25-OH-D_3 in pharmaceutical products. The conditions used were the same as those described above for the parent vitamin and the oxidation peak again occurred at +1.1 V; this was suitable for analytical purposes. A second voltammetric peak appeared at +1.3 V but this was

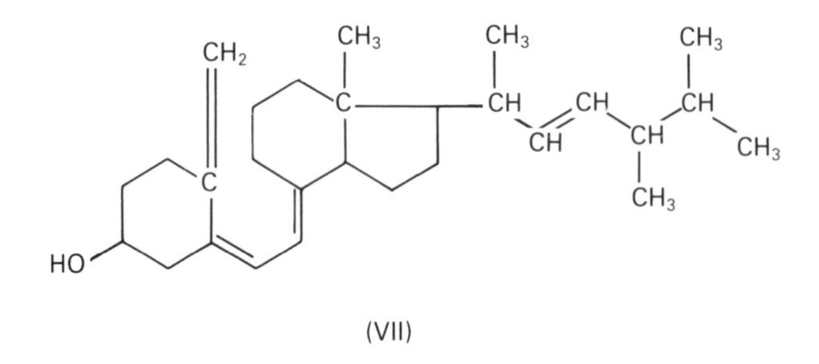

(VII)

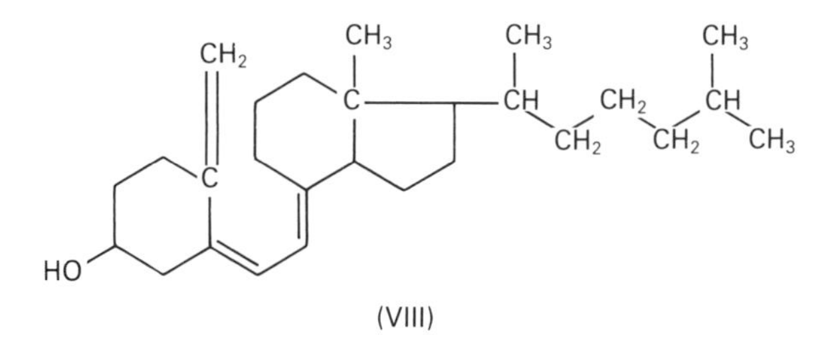

(VIII)

poorly defined and was not analytically useful. The same group [24] have also developed methods for the determination of vitamin D_3 and 25-OH-D_3 individually in pharmaceutical products using flow-injection analysis; the amperometric detector contained a glassy carbon working electrode. The carrier solution consisted of 60% methanol–0.075-M lithium perchlorate and the detector was set at a potential of +1.05 V for both substances. The detection limits were found to be 7 ng and 11 ng for vitamin D_3 and 25-OH-D_3 respectively.

4.2.3 Vitamin E

4.2.3.1 Methods involving polarographic and voltammetric techniques

Vitamin E comprises a group of structurally similar compounds known as the tocopherols; the most common of these are α-, β-, γ- and δ-tocopherol, which have the following structures:
where $R_1=CH_3$, $R_2=CH_3$, $R_3=CH_3$ in α-tocopherol (IX); $R_1=CH_3$, $R_2=H$, $R_3=CH_3$ in β-tocopherol (X); $R_1=H$, $R_2=CH_3$, $R_3=CH_3$ in γ-tocopherol (XI); $R_1=H$, $R_2=H$, $R_1=CH_3$ in δ-tocopherol (XII).

The electrochemical activity of the E vitamers is associated with the phenolic hydroxy group, which readily undergoes oxidation.

Polarography has been employed for the determination of α-tocopherol in the presence of the β- and γ-vitamers; it was found that α-tocopherol was the easiest to oxidize, which allowed its determination in mixtures containing the three vitamin E forms [25,26]. A polarographic method has also been developed for the determi-

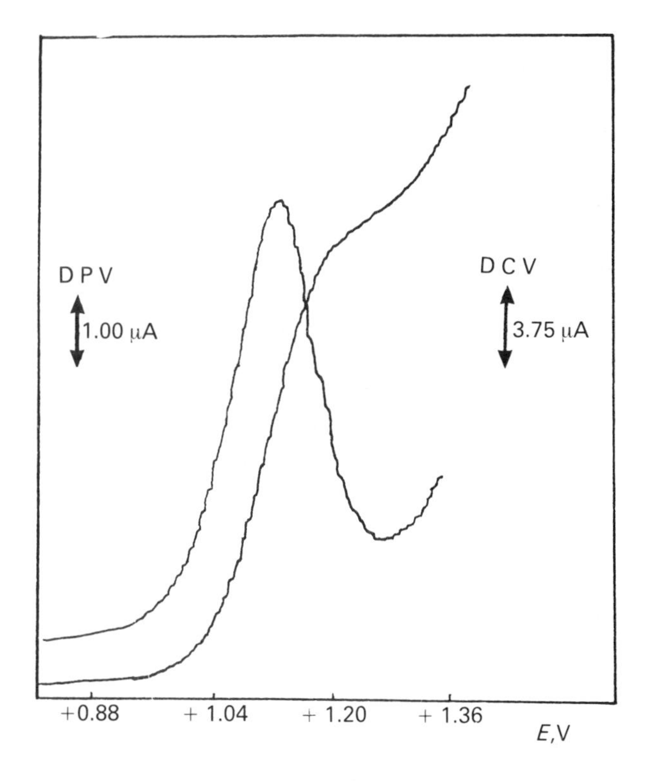

Fig. 4.5 — Voltammetric behaviour of vitamin D_3 at a glassy carbon electrode; concentration of vitamin D_3, 1.3×10^{-4} M; concentration of $LiClO_4$, 0.075 M in methanol. (Reproduced from [22] by permission of the copyright holders.)

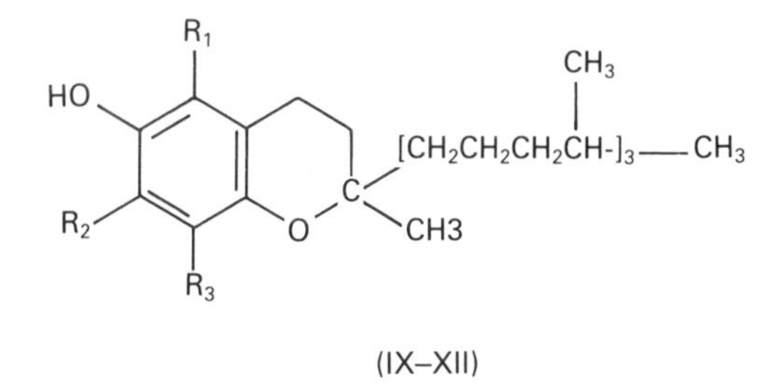

nation of the vitamin E metabolite tocopheronolactone [27]; the method was applied to urine samples following a one gram dose of α-tocopherol. The purified extract was dissolved in ethanolic acetate buffer pH 5.3 and the determination was made using the resulting reduction wave with $E_{1/2}=-0.1$ V. A similar procedure, for the same metabolite, has been described which incorporated a thin-layer chromatographic step [28].

Voltammetric methods employing carbon electrodes have become increasingly popular for determinations of vitamin E. McBride and Evans [29] used a glassy carbon electrode and a linear sweep waveform for the compounds (IX–XII). In a supporting electrolyte containing 0.12 M sulphuric acid in ethanol + benzene (2+1) the peak potentials were found to be +0.57, +0.67 and +0.74 V (versus SCE) for (IX), (X) and both (XI) and (XII). The method was applied directly to the analysis of five different oils. In a similar study Atuma and Lindquist [30] preferred to use carbon paste as the working electrode for the determination of tocopherols in vegetable oils, foods and pharmaceuticals; in this study, saponification of the samples was performed to remove interferences prior to the voltammetric determinations.

Improved detection limits for the voltammetric detection of tocopherols was reported by Loeliger and Saucy [31]; these authors employed a micro flow-through cell containing a glassy carbon working electrode. Concentrations down to 0.01 μg cm^{-3} could be measured and the method was used to determine the E vitamers in samples such as maize oil, butter and carrots. A method has been reported for the measurement of tocopherols in plant oils using a glassy carbon electrode [32].

Brieskorn and Mahlmeister [33] have used a rotating disc glassy carbon electrode, with a differential pulse waveform, for the determination of α-tocopherol; it was reported that with a rotation speed of 70 rpm the magnitude of the signal increased by over 100% when compared with the static electrode. In this study, the authors applied the method to the determination of the same vitamer in human serum. Following a suitable sample preparation procedure the extract containing vitamin E was dissolved in ethanol containing 0.1-M sulphuric acid. The recovery of the method was found to be in the range 96.1 to 102.0%. Shiozaki *et al.* [34] also used a glassy carbon electrode for the analysis of vitamin E; however, in this case the determination was carried out in a non-aqueous supporting electrolyte consisting of 0.5-M $LiClO_4$ dissolved in acetonitrile.

A selective AdSV method for the determination of vitamin E was reported by Wang and Freiha [35]. In this method, the vitamin was preconcentrated into a carbon paste electrode at open circuit with stirring; the electrode was then transferred to a fresh solution containing phosphate buffer for the measurement step using DPV. This preconcentration/medium-exchange procedure was applied to the determination of vitamin E in a multivitamin preparation and corn oil sample; Fig. 4.6 clearly demonstrates the feasibility of measuring the vitamin in these samples.

Several reviews have appeared which discussed electrochemical mechanisms and voltammetric methods for the analysis for vitamin E [21,25,36].

4.2.3.2 Methods involving LCEC

The circulating blood levels of tocopherols have been of clinical interest because a deficiency can cause neuropathological and neuromuscular disorders in children; one possible cause of difficiency is malabsorption [37,38]. Studies of vitamin E levels might also throw some light on the recommended dietary allowances (RDA) for the elderly; there still seems to be some conflict as to what is a safe and adequate daily

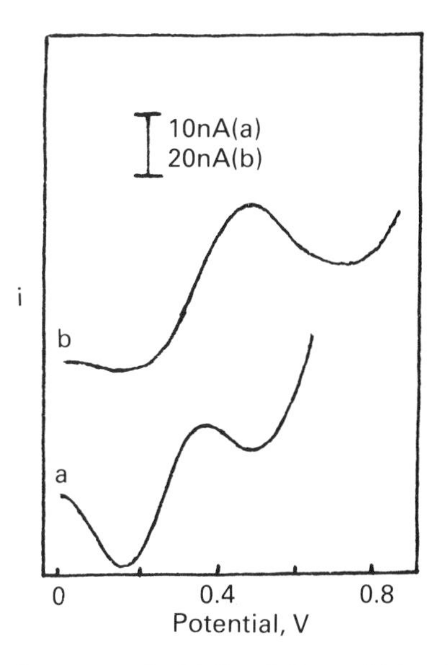

Fig. 4.6 — Differential-pulse response for (a) One-A-Day multivitamin and (b) Mazola corn oil samples. Preconcentration for 2 min at 550 rpm stirring; measurement after medium exchange to a phosphate buffer solution. Carbon (silicone grease) paste electrode. The multivitamin tablet and 1 g of corn oil were diluted in 100 cm^3 of ethanol; 2 cm^3 of the multivitamin preparation solution or 1 cm^3 of the oil solution was dissolved with phosphate buffer to 10 cm^3 to prepare the sample. Exchange solution, 0.1-M phosphate buffer. (Reproduced from [35] by permission of the copyright holders, Elsevier Science Publishers Physical Sciences and Engineering Division.)

intake of this vitamin [39]. A very recent, and interesting, article indicated that vitamin E 'mops up' some of the free radicals that are formed in the body [40]; since free radicals have been implicated in many diseased states, including cancer and rheumatoid arthritis, investigations on E vitamers in the blood and other biological fluids might be an area of considerable importance. For these reasons it is clear that reliable analytical methods for the assay of this vitamin are desirable.

The oxidation of the phenolic hydroxy group, which is present in all of the E vitamers, can be exploited using LCEC. Since δ-tocopherol does not appear to circulate at measurable concentrations it may be conveniently used as an internal standard. Chou *et al.* [41] reported on an LCEC method for the determination of α-tocopherol, and total β- plus γ-tocopherol, in only 25 μl of serum. The vitamers were extracted by the addition of 100 μl of ethanol and 100 μl of heptane to the serum sample. A 50-μl aliquot of the heptane phase was evaporated to dryness and the residue was reconstituted in 50 μl of methanol for injection onto an ODS column; the mobile phase consisted of 95% methanol–acetate buffer pH 5.0. Amperometric detection was performed with a thin-layer cell containing a glassy carbon working electrode; this was operated in the oxidative mode at a potential of +1.0 V. The LCEC chromatograms recorded on the serum extracts were surprisingly free of interferences considering the minimal clean-up used (Fig. 4.7).

The predominant form of the vitamin in normal subjects was found to be

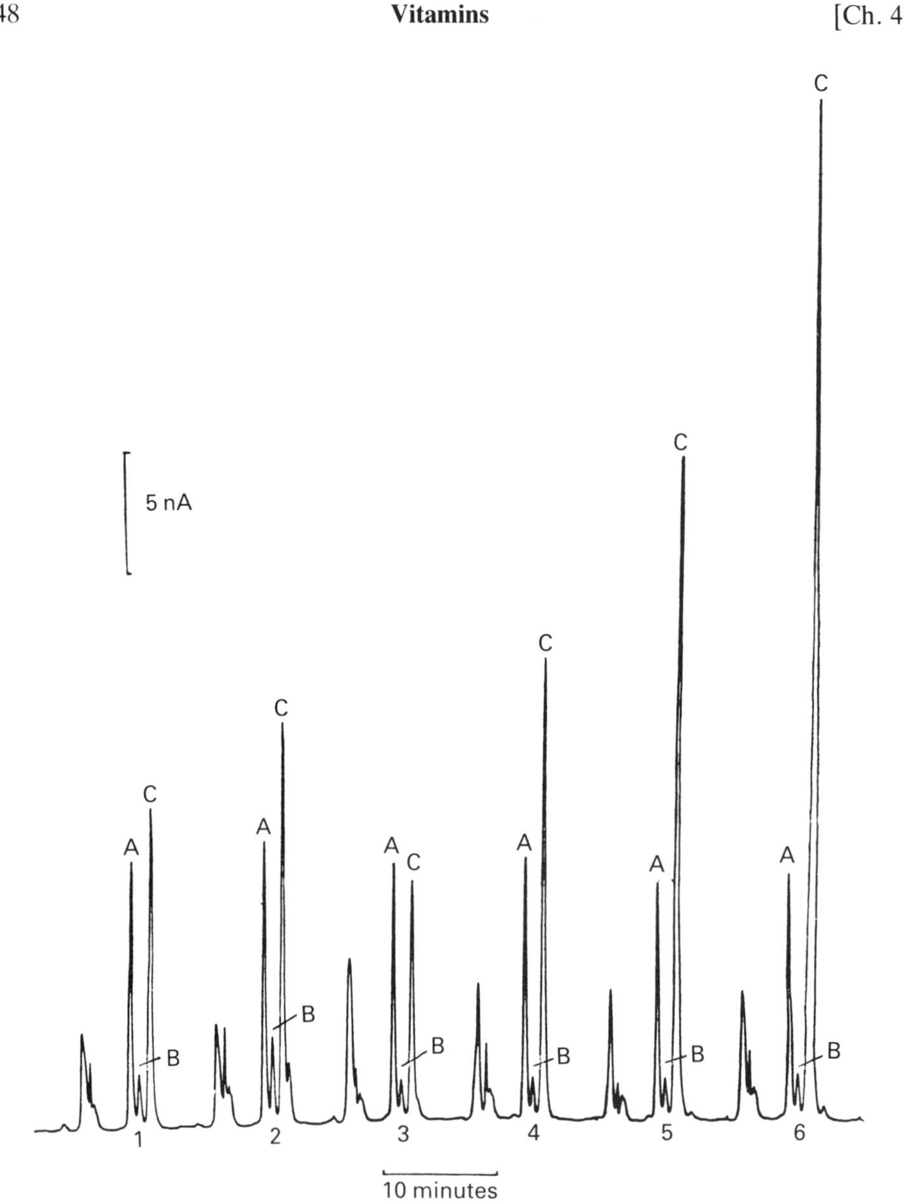

Fig. 4.7 — Typical chromatograms obtained for six different sera. (A) δ-tocopherol; (C) α-tocopherol. Both β- and γ-tocopherol are co-eluted as one component (B). Sample 5 contains moderately increased α-tocopherol (45.6 mg dm^{-3}); sample 6 has a very high α-tocopherol concentration (69.9 mg dm^{-3}). (Reprinted with permission from *Clinical Chemistry*, (1985) Vol. 31, Page 880, Figure 3. Copyright American Association for Clinical Chemistry, Inc.)

α-tocopherol; the limit of detection for the three vitamers was below 0.1 $\mu g\ cm^{-3}$, which was well below the normal circulating concentrations (4.3–9.7 $\mu g\ cm^{-3}$ for α-tocopherol; 1.8–3.9 $\mu g\ cm^{-3}$ for β- and γ-tocopherol). The authors considered that the use of electrochemical detection had facilitated the applicability of the method to the neonatal population because only 25 μl of sample were required [42].

A similar LCEC assay for vitamin E has been described by Vandewoude *et al.* [43]. After separation of the tocopherols from 100 μl of plasma the resulting residue was dissolved in isopropanol for injection onto the reversed-phase column. In this case the mobile phase consisted of methanol–0.1% pyridine–0.05-M sodium perchlorate; the potential applied to the glassy carbon electrode was +0.7 V. The method compared well with an HPLC method involving UV detection and was apparently less susceptible to interference from co-extracted material

Castle and Cooke [44] have developed a simple extraction procedure for α-tocopherol from serum, and plasma, prior to LCEC. It was necessary only to perform protein precipitation and after centrifugation a 10-μl (or 5-μl) aliquot of supernatant was injected onto the reversed-phase column (45 mm × 4.6 mm i.d.). This approach was possible because a selective dual-electrode detection system was employed; this consisted of two porous graphite electrodes in series which were operated in the 'screen mode'. The upstream electrode was held at +180 mV and the downstream electrode (detector) was set at +500 mV. The simple extraction procedure, combined with short elution time, resulted in the analysis of up to 10 samples/h; in addition the assay could be performed with very small volumes of serum or plasma (less than 50 μl).

The simultaneous determination of α-tocopherol, retinol and β-carotene in serum has been reported by MacCrehan and Schonberger [45]. These authors found that it was necessary to use a slow gradient elution to resolve the three analytes from their isomers and from other constitutents of serum. It was also recommended that an internal standard (tocol) should be used because of incomplete separation of hexane from serum; this would have resulted in recoveries of less than 100%. In addition, the workers showed that it was beneficial to use a three-component solvent system; water/methanol/*n*-butanol; the supporting electrolyte was ammonium acetate pH 3.5. For the simultaneous determination of the three analytes the gradient was as follows: initial solvent, water/methanol/*n*-butanol (15:75:10, by vol); final solvent, water/methanol/*n*-butanol (2:88:10, by vol). The initial solvent composition was used for 3 min, then was changed linearly to the final solvent composition over a 15-min period; this was held for 17 min before being returned to the initial solvent composition. Fig. 4.8 shows the resulting LCEC chromatogram together with the LCUV chromatogram for comparison. It was shown that the detection limits were superior with the electrochemical detector; however, the selectivity of the UV detector was superior for β-carotene, particularly when retinyl palmitate was present.

The simultaneous determination of tocopherols, and ubiquinols in blood, plasma, tissue homogenates and subcellular fractions has been described by Lang *et al.* [46,47]. Extraction was performed by the method of Burton *et al.* [48], which was found to be superior to other methods in terms of simplicity and speed. The mobile phase consisted of methanol–ethanol (1:9) and 20-mM lithium perchlorate; the flow-through cell contained a glassy carbon working electrode which was operated in the oxidative mode. The separation and detection of δ-, γ- and α-tocopherol in human blood and plasma was possible.

Another example of the application of LCEC to the determination of vitamin E

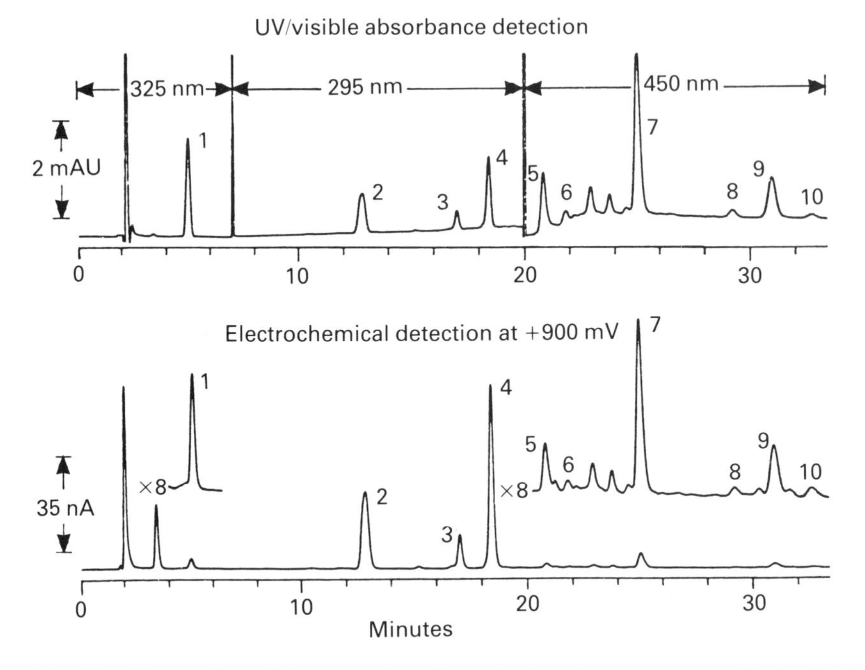

Fig. 4.8 — Slow-gradient-elution separation of micronutrients in a serum extract. Peaks: (1) all-*trans*-retinol; (2) tocol; (3) γ-tocopherol; (4) α-tocopherol; (5) lutein; (6) zeaxanthin; (7) cryptoxanthin; (8) α-carotene; (9) all-*trans*-carotene; (10) cis-β-carotene. (Reprinted with permission from *Clinical Chemistry* (1987) Vol. 33, Page 1585, Figure 2. Copyright American Association for Clinical Chemistry, Inc.)

in small biological samples has been described by Pascoe *et al.* [49]; in this case the method was designed for the simultaneous measurement of both α-tocopherol and α-tocopherolquinone. This was feasible by the utilization of dual glassy carbon electrodes, in series, operated in the redox mode; the upstream electrode (generator) was held at −0.70 V and the downstream electrode (detector) was set at +0.6 V.

Another application of LCEC for the determination of vitamin E was in samples of animal feeds [50].

A novel concept for the separation and detection of α-tocopherol, using normal phase LCEC, has been reported by Schieffer [51]. The electrochemical cell consisted of an annular porous working electrode; this was fabricated from crushed RVC=reticulated vitreous carbon, an outer cation exchange membrane and an inner Teflon tube. The mobile phase consisted of hexane-1-propanol (93:7) without supporting electrolyte. The authors indicated that further work was required to optimize cell geometry because some loss of response occurred with time.

4.2.4 Vitamin K

4.2.4.1 Methods involving polarography and voltammetry

The group of compounds known as vitamin K are structurally similar and will, therefore, be discussed together.

Vitamin K_1 (XIII, phylloquinone), K_2 (XIV, menaquinone group) and K_3 (XV, menadione) have the structures shown below

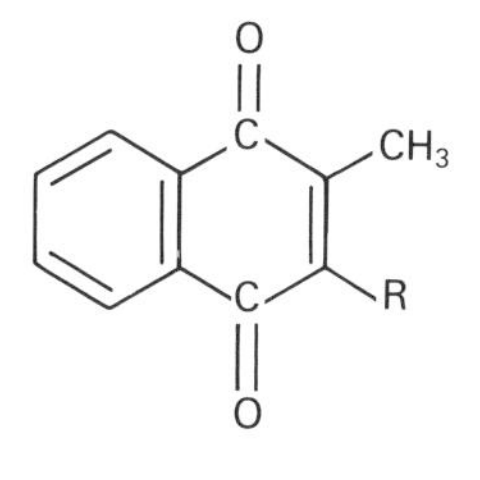

(XIII–XV)

where R=H in menadione,

$$R = \text{-}CH_2\text{-}CH{=}\underset{}{\overset{CH_3}{\overset{|}{C}}}\text{-}CH_2\text{-}[\text{-}CH_2\text{-}CH_2\text{-}\overset{CH_3}{\overset{|}{C}}H\text{-}CH_2\text{-}]_3\text{-}H$$

in phylloquinone

$$R = \text{-}[\text{-}CH_2-CH{=}\overset{CH_3}{\overset{|}{C}}\text{-}CH_2\text{-}]_n\text{-}H \quad (n=1\text{–}13)$$

in menaquinones (MK-n)

These compounds all contain the 1,4-naphthoquinone moiety; therefore, similar electrochemical characteristics might be expected.

Investigations into the electrochemical behaviour of the K vitamers have been of considerable interest to several research groups because of the possibility of exploiting this behaviour for their sensitive measurement in a variety of matrices. Therefore, the first part of this subsection briefly discusses the electrochemical behaviour of the major forms of vitamin K; this is followed by examples describing the application of polarographic and voltammetric methods for their analysis.

The author [4] has carried out studies on the electrochemical reduction of vitamin K_1 over a wide pH range using sampled d.c. polarography; the supporting electrolytes were prepared from Britton–Robinson buffers and contained 90% ethanol to completely solubilize the vitamin. Under these conditions one well-defined, pH-dependent, cathodic wave was obtained which moved to more negative potentials with increasing pH. The $E_{1/2}$ was found to obey the relationship

$$E_{1/2}(\text{V}) = +0.1 - 0.06\,\text{pH} \tag{4.2}$$

The overall number of electrons (n) transferred per molecule, throughout the pH range studied, was found to be two. The number of protons (m) involved in the reduction process was determined from the slope of the $E_{1/2}$ versus pH plot and was calculated to be two [1] (see section 1.1.1.1).

The electrochemical behaviour of phylloquinone (XIII) has also been studied by Vire and Patriarche [52, 53]; these investigators employed d.c, a.c. and differential-pulse polarography to aid in the elucidation of the reduction mechanism. In this case, the vitamin was dissolved in solutions containing sodium acetate–acetic acid/methanol (15:85). In acid or neutral solutions, only one well-defined peak or wave was observed; however, when the pH of the supporting electrolyte was 9.6 both a.c. and differential-pulse polarograms exhibited two peaks. This was considered to be the result of adsorption of vitamin K_1 at the electrode surface.

Detailed studies were carried out by Hart and Catterall [54] to investigate the possibility that phylloquinone could undergo adsorption at mercury electrodes; the technique used for this purpose was cyclic voltammetry with an HMDE. The vitamin (4 μM) was dissolved in acetate buffers (pH 6.0) containing 70% ethanol, 90% ethanol, 70% methanol, 90% methanol; the scan rates were between 10 and 100 mV s^{-1}. The cyclic voltammograms obtained on these four solutions using a scan rate of 100 mv/s is shown in Fig. 4.9; the presence of adsoprtion is clearly indicated by the triangular symmetry of the peaks obtained for the vitamin dissolved in 70% alcoholic solutions. The magnitude of the peak currents indicated that adsorption was greatest in the 70% methanolic solution. It should be added that the cyclic voltammograms also indicated that the reduction was reversible and that the reduced form of the vitamin was also adsorbed.

Further evidence for the presence of adsorption was obtained by plotting $i_p/c\upsilon^{1/2}$ vs $\upsilon^{1/2}$(where i_p is the peak current (μA), c is the vitamin concentration (mmol dm^{-3}), and υ is the scan rate (mV s^{-1})). All of these graphs showed positive slopes, indicating the presence of adsorption; however, for solutions containing 90% ethyl alcohol, adsorption appeared to be quite weak.

The investigations discussed above strongly suggest that vitamin K_1 undergoes a reversible $2e^-$, $2H^+$ reduction to produce the corresponding hydroquinone (Fig. 4.10). In addition, the vitamin has also been shown to undergo adsorption at the mercury electrode. The result reported [52–54] suggest that the extent of adsorption is dependent on several factors; these include the concentration and type of alcohol used in the supporting electrolyte and the pH of the buffer. This phenomenon has been exploited to increase the sensitivity of voltammetric methods and is discussed later in this section.

The electrochemical behaviour of a vitamin K_2 compound (menaquinone-4) was studied by Takamura and Hayakawa [55] using d.c. and a.c. polarography and CV at an HMDE. These authors reported that only one diffusion-controlled d.c. polarographic wave was observed in a solution containing 1.00×10^{-4} M of the vitamin dissolved in 70% methanolic acetate buffer pH 4.8; these solutions also contained 0.2-M sodium perchlorate. The proposed mechanism for the reduction of the naphthoquinone moiety was the same as that shown in Fig. 4.10 for vitamin K_1, i.e. a two-electron, two-proton reversible reduction to produce the corresponding hydroquinone. In addition, it was reported that the a.c. polarograms showed three

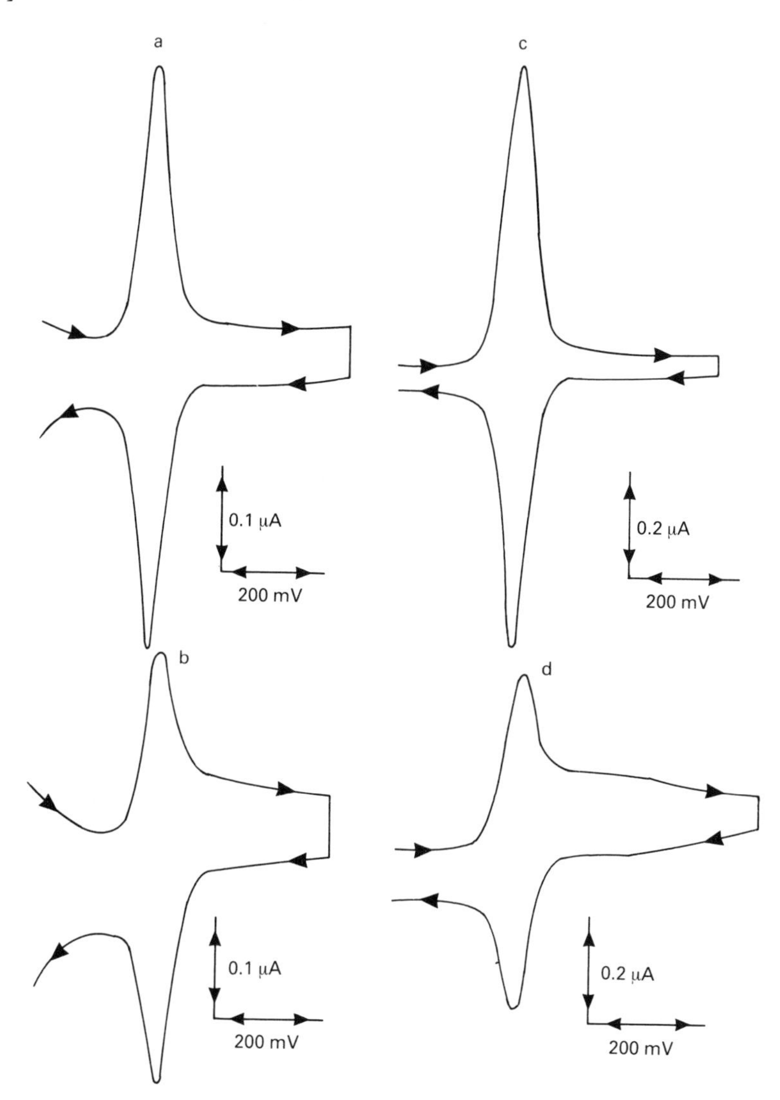

Fig. 4.9 — Cyclic voltammograms for acetate-buffered (pH 6.0) 4-μM vitamin K_1 solutions; (a) 70% ethanol, (b) 90% ethanol, (c) 70% methanol, (d) 90% methanol. Initial potential -0.1 V, scan rate 100 mV s^{-1}. (Reproduced from [54] by permission of the copyright holders, Elsevier Science Publishers Physical Sciences and Engineering Division.)

cathodic peaks, two of which were associated with adsorption behaviour. Further evidence for the presence of adsorption was obtained from electrocapillary data using plots of mercury drop-time versus applied potential; these showed an inflection in the region 0 V to -0.5 V in the presence of the vitamin but this was absent in supporting electrolyte only. Conclusive evidence for adsorption was then obtained by the CV technique where similar studies to those described above for phylloqui-

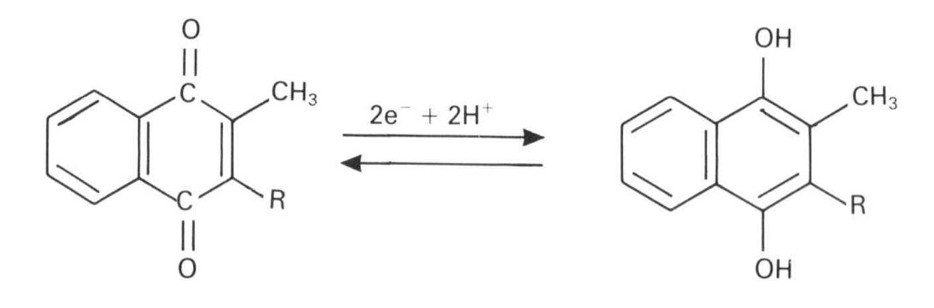

Fig. 4.10 — Mechanism of reduction of vitamin K_1.

none were carried out. It was concluded that menaquinone-4 was adsorbed at mercury via the long isoprenoid side-chain.

The nature of the electrochemical reduction process for vitamin K_3 has been investigated by Patriarche and Lingane [56] using d.c. polarography and CV at an HMDE; the supporting electrolytes consisted of phosphate, or acetate buffers, containing between 3–10% methanol. The number of electrons involved in the reduction process was found to be two; the product of reduction was the corresponding hydroquinone. This process was shown to be quasi-reversible because the difference between the cathodic and anodic CV peaks was greater than the theoretical value (i.e. greater than $59/n$ mV). Subsequent electrochemical studies by Patriarche and Vire [52,53] on this vitamin indicated that adsorption occurred at the mercury electrode at pH 7. However, Takamura and Hayakawa [55] carried out CV studies on the same vitamin in 70% methanolic acetate buffer pH 4.8, and reported that the reduction was simply diffusion-controlled with no indication of adsorption. Therefore, it would appear that concentration of alcohol, and pH of the buffer, may play an important role in the adsorption of vitamin K_3 at mercury; this was also found for phylloquinone and was discussed earlier.

The electrochemical behaviour of vitamins K_1 and K_3 has also been investigated at carbon electrodes. Hart *et al.* [57] performed cyclic voltammetry on vitamin K_1 using a glassy carbon electrode in a supporting electrolyte containing 95% MeOH–0.05-M acetate buffer pH 3.0. One cathodic peak was obtained on the negative scan ($E_p = -0.48$ V), which is consistent with a $2e^-$, $2H^+$ reduction on the quinone to the hydroquinone; the reverse scan also showed one anodic peak ($E_p = 0$ V), which was due to reoxidation of the vitamin back to the quinone. However, the separation between the cathodic and anodic peaks was clearly greater than $59/n$ mV, which indicates that the process is quasi-reversible. In addition, the product, but not the reactant, was found to undergo adsorption at the glassy carbon electrode. In contrast, phylloquinone was reported to produce two reduction peaks using thin-layer voltammetry at a carbon electrode [58]. It was shown that the first peak was due to a one-electron, one-proton reduction to form the semi-quinone and the second to a further one-electron, one-proton transfer producing the hydroquinone. However, in the same study the authors found that menadione was reduced in a reversible $2e^-$, $2H^+$ reaction to the hydroquinone.

Cauquis and Marbach [59] have investigated the electrochemical reactions occurring for menadione (Q) at a platinum electrode in acetonitrile containing 0.1-M tetraethylammonium perchlorate. In neutral, non-buffered solution, CV showed two peaks on both the forward and the reverse scans; this indicated a stepwise reduction to the dianion (Q^{2-}). In the presence of water, a different mechanism was found to operate and the monoanion, QH^-, was produced. In the same investigation the authors reported that in the presence of strong acids, reduction proceeded to the hydroquinone, QH_2.

The nature of the electrochemical reduction for menadione bisulphite (XVI), which is a synthetic vitamin K_3 compound, has been studied by Vire and Patriarche [60]. It was shown that the reduction mechanism was different to that of menadione itself this difference is due to the loss of conjugation between the 1, 2, 3, and 4 positions. The reduction process of the bisulphite derivative was found to be pH-dependent and was rather complex. In neutral solution (pH 7) two d.c. waves were reported and the mechanism of reduction was that shown in Fig. 4.11. In strong acid

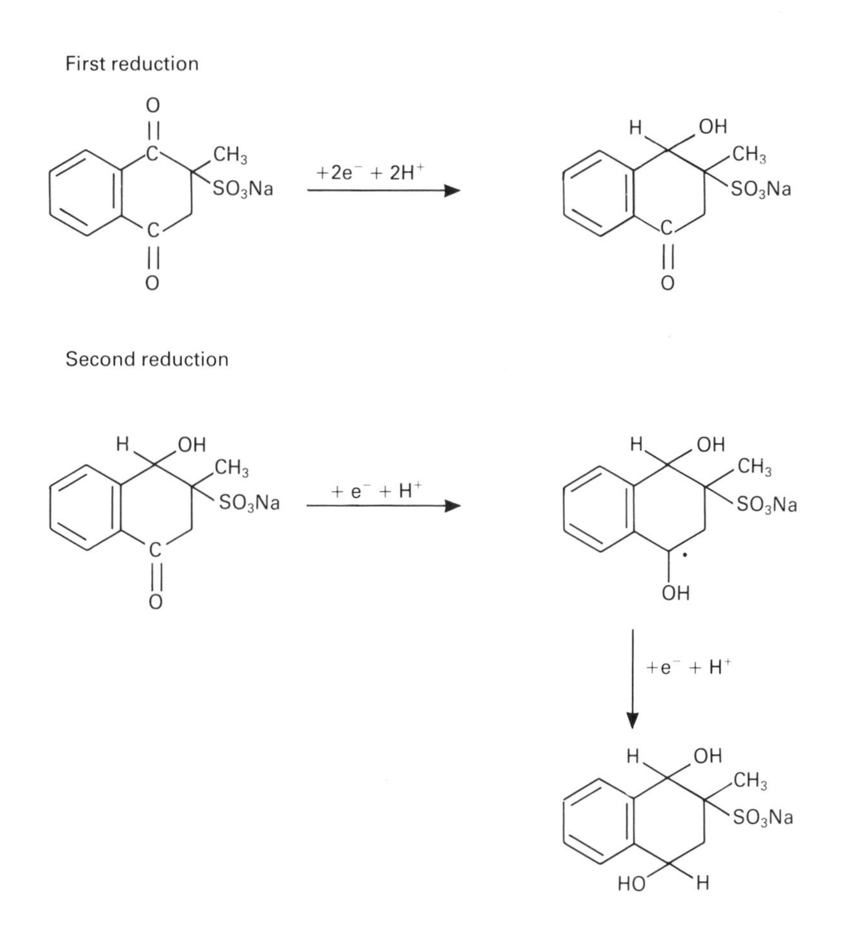

Fig. 4.11 — Mechanism of reduction of bisulphite derivative.

solution the second wave was found to merge with the first, which resulted in one wave with increased magnitude. At high pH values the polarographic behaviour became even more complex owing to the degradation of the compound to produce menadione; this occurred by simple cleavage at the 2-position.

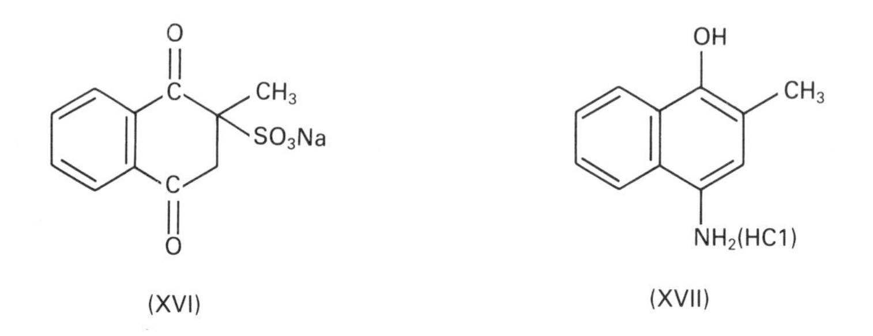

A detailed study of the degradation of menadione bisulphite was carried out by Vire *et al.* [61] using d.c., a.c. and differential-pulse polarography. In this investigation at total of eight different polarographic signals were observed which corresponded to the reduction of eight different compounds.

The electrochemical oxidation of vitamin K_5 [XVII] (4-amino-2-methyl-1-naphthol hydrochloride) was investigated by Takamura and Watanabe [62] using d.c. and a.c. polarography. The d.c. polarogram obtained at pH 3.3 showed two well-defined anodic waves; the first was stated to be the result of oxidation to produce the corresponding quinoimine (Fig. 4.12). The second anodic wave was apparently due

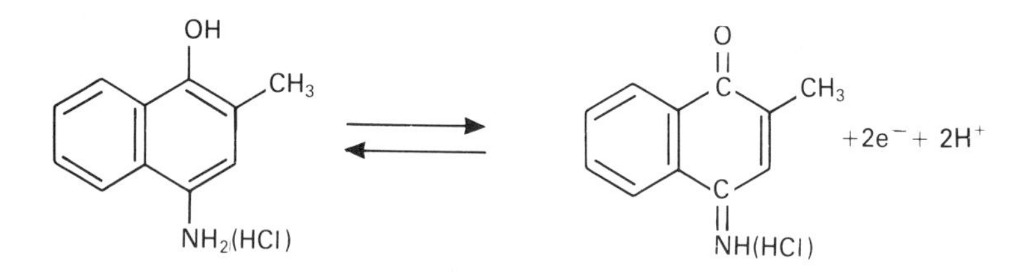

Fig. 4.12 — Mechanism of reduction of vitamin K_5.

to the formation of mercurous chloride, which formed a deposit on the mercury electrode; the chloride was supplied through dissociation of the vitamin.

Since DPP can offer, under favourable conditions, limits of detection of about 10^{-8} M it has been applied in several investigations to measure K vitamins.

Hart and Catterall [63] have carried out detailed studies to find the optimum conditions for the determination of low levels of phylloquinone by DPP. In this investigation the effect of ionic strength, concentration of ethanol and the pH of acetate buffers on the DPP peaks was examined. It was shown that the peak current, measured at −0.360 V versus the SCE was proportional to the concentration of phylloquinone over the range 0.26 μM to 1.5 μM when the supporting electrolyte contained 90% ethanol–0.05 M acetate buffer pH 6.0. Hart and coworkers [64] subsequently employed this technique to follow the clearance rates of the vitamin in normal subjects and patients with bone fractures, after a 20-mg intravenous injection. For the determination in plasma, extraction was performed with chloroform and methanol; following chloroform evaporation, the residue was dissolved in the supporting electrolyte and subjected to DPP. The polarogram exhibited a peak at about −0.58 V versus SCE, which was due to the reduction of the vitamin. The shift in the reduction potential, compared to buffer only, was considered to be an effect of co-extracted lipid material. However, this did not interfere with the quantitative determination of the vitamin in plasma extracts; the sensitivity of the method was suitable for the assay of concentrations down to 200 ng/ml of plasma. There was no interference from other naturally occurring substances (Fig. 4.13). The results of these studies supported the hypothesis that vitamin K may be involved in a metabolic cycle which underlies calcification of bone.

Other workers have also explored the possibility of using DPP for the measurement of low concentrations of phylloquinone, although they did not actually apply the technique to the analysis of real samples [52, 65].

DPP has also been applied to the determination of menadione and menadione bisulphite in plasma; these studies were primarily concerned with determining the stability of these vitamins in plasma [1]. It was shown that the measurement of the bisulphite derivative in plasma samples was feasible when using 0.5 M H_2SO_4 as a supporting electrolyte; it was also necessary to add *n*-octanol to prevent excess foaming during deaeration. Under these conditions, a well-defined DPP peak occurred at −0.64 V versus SCE; concentrations down to 230 ng cm^{-3} plasma could be easily measured. Using this procedure, it was shown that menadione bisulphite, added to plasma samples, quickly degraded and the degradation product could not be polarographically detected at the potentials employed. Further studies revealed that the product of degradation was menadione itself, which was formed by simple cleavage of the bisulphite group. In order to monitor accurately plasma menadione concentrations a DPP method was developed that involved a preliminary extraction with peroxide free diethyl ether. The resulting residue was dissolved in a supporting electrolyte comprising 90% ethanol–0.05 M acetate buffer pH 3.0. A well-defined peak was observed at a potential of −0.5 V versus SCE and the recovery of menadione was found to be 90%.

In a related study, it was of interest to ascertain whether menadione existed in the 'free' state in plasma after storing at 37°C [1]. Plasma samples were spiked with menadione and incubated for various times prior to assaying by the modified DPP method. It was discovered that the peak at −0.50 V decreased rapidly and the $T_{1/2}$ value was calculated to be about 28 min. These resulta were in agreement with those of Scudi and Buhs [66] who used a colorimetric method involving cysteine and showed that free menadione disappeared rapidly from plasma; these authors indicated that menadione reacted with plasma proteins. Therefore, in these polarographic studies it would seem that either the menadione–plasma complex was not polarographically active, or that this complex was not extracted into ether.

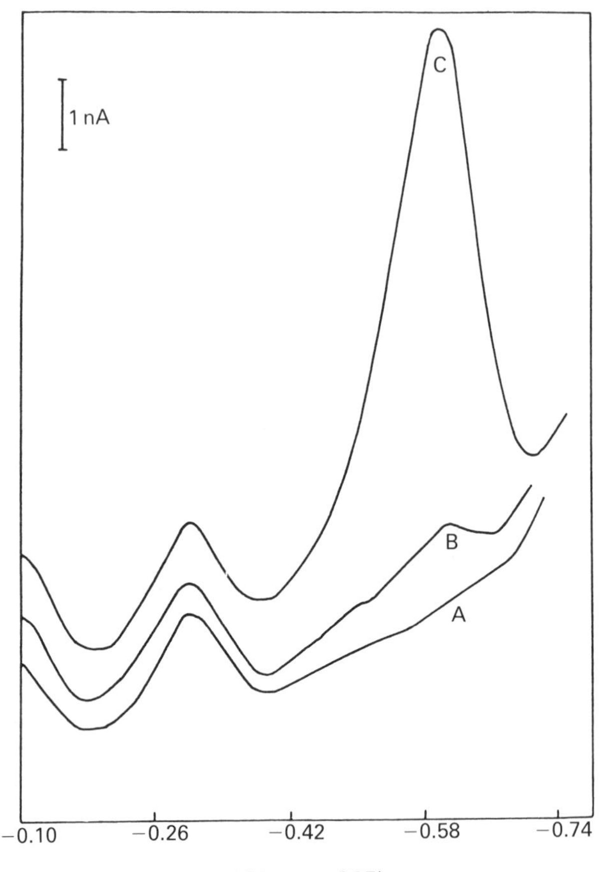

Fig. 4.13 — Differential pulse polarograms of (a) control plasma, (b) plasma spiked with 412 ng of vitamin K_1, (c) plasma spiked with 6.18 μg of vitamin K_1. (Reproduced from [64] by permission of the copyright holders, Elsevier Science Publishers, Physical Sciences and Engineering Division.)

In contrast, it appears that a polarographic method has been developed that is capable of measuring menadione in spiked plasma samples and in plasma following an oral dose of menadiol diphosphate [67]. In this study, a solvent extraction step was carried out with diethyl ether, and the plasma phase was frozen before transferring the ether to a separate vessel. This procedure was repeated and the ether extracts were added together; to the combined extracts was added calcium chloride. After separating off the calcium chloride the ether was evaporated to dryness and the residue dissolved in buffer; this was then analysed by DPP and a peak for menadione was obtained at −0.28 V. This peak was then used to determine the plasma levels of the vitamin following oral doses. It would be of interest to evaluate this method on spiked plasma samples incubated at 37°C for various times as discussed above; this may throw some light on the possible menadione–protein interactions previously observed [1,66].

Stationary electrodes have also been investigated for application in the voltammetric determination of K vitamins.

In one study DPV at an HMDE indicated that this approach was more sensitive than DPP for the analysis of phylloquinone [54]. The limit of detection for K_1 was 10 ng/ml in a supporting electrolyte containing 60% methanolic acetate buffer; this increased sensitivity appeared to be the result of adsorption of the vitamin onto the HMDE. However, it should be added that these determinations were performed in supporting electrolyte only. In order to apply this to the measurement of phylloquinone in plasma it would be necessary to purify plasma extracts after the solvent extraction step.

This adsorption behaviour of vitamin K_1 has been exploited by Vire *et al.* [68] for its determination at low levels by AdSV at an HMDE. The application of differential pulse and square waveforms were investigated and the latter was found to give the more sensitive signal (Fig. 4.14).

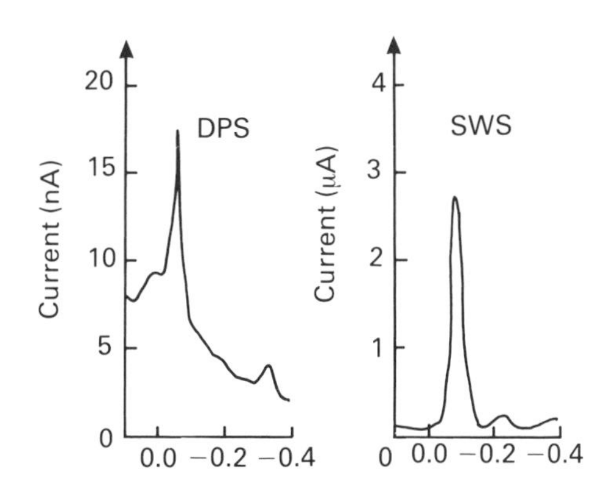

Fig. 4.14 — Comparison of the voltammetric response of differential-pulse stripping voltammetry (DPSV) and square-wave stripping voltammetry (SWSV) Vitamin K_1, 4×10^{-7} M; methanol 5%; pH 4.2; accumulation time 300 s; scan rate DPSV 2 mV s^{-1}, SWSV 200 mV s^{-1}; pulse height DPSV and SWSV 20 mV; frequency SWSV, 100 Hz. (Reproduced from [68] by permission of the copyright holders, Marcel Dekker.)

The optimum supporting electrolyte consisted of 5% MeOH–K_2SO_4/H_2SO_4, pH 4.2 (this solution contained a salt concentration of 3×10^{-2} M); the best deposition potential was +0.1 V. When a deposition time of 300 s was used, together with blank subtraction, it was possible to obtain linear calibration graphs over the range 1×10^{-9} to 1×10^{-7} M.

In a separate study, the same research group [69] have shown that vitamin K_3 undergoes adsorption at the HMDE. In this case, it was found that the reduced form of the vitamin was more strongly adsorbed than the quinone form and the most suitable supporting electrolyte was 0.3-M perchloric acid; consequently, AdSV using a square waveform could be used to enhance the voltammetric signal. Interestingly, the currents obtained for menadione were found to be about 10 times greater than those obtained for phylloquinone; this difference was considered to be the result of steric hindrance by the phytyl chain of the latter compound. The limit of detection for vitamin K_3 was reported to be 1.3×10^{-10} M.

Another approach, utilizing preconcentration/voltammetric determination of vitamin K_1, has been investigated by Hart *et al.* [70]. In this study, the possibility of accumulating the vitamin at carbon paste electrodes, prepared from Nujol, with a variety of graphite powders was examined. It was shown that a mixture of Nujol/ultra-superior purity (USP) graphite (25/75 m/m) gave the greatest sensitivity with AdSV (Fig. 4.15); the optimum accumulation time was found to be 15 min at open circuit.

In the same study [70] a variety of procedures were investigated to separate the vitamin from plasma prior to adsorptive stripping analysis. A solvent extraction method using hexane and ethanol gave the best recovery (91%) and detection limit (180 ng cm^{-3} in the supporting electrolyte; 450 ng cm^{-3} of plasma). However, the analysis time could be reduced by 50% (with some loss in sensitivity) by using ethanol to precipitate the proteins and with the measurement being made directly on the resulting supernatant. Since calibration graphs are linear, quantitation may be performed by the method of single standard addition; therefore, relatively short analysis times are feasible.

4.2.4.2 *Methods involving LCEC*

The search for a sensitive, selective analytical method for the determination of normal and subnormal endogenous levels of vitamin K compounds has been of interest to workers in a variety of biomedical fields. However these determinations are difficult because of the low levels present and also because interference may occur from other plasma compounds.

One of the first studies to indicate that LCEC may be an appropriate technique for such analysis was reported by Ikenoya *et al.* [71]. In this study phylloquinone was extracted from rat plasma, into hexane, following a 1 mg/kg oral dose; the residue from the solvent layer was then dissolved in isopropyl alcohol for injection onto an octadecylsilyl reversed-phase column. The mobile phase consisted of $NaClO_4$–$HClO_4$ dissolved in ethanol–methanol (6:4); detection was performed at a potential of −0.3 V (versus Ag/AgCl). The authors stated that the minimum detection limit was 100 pg on column.

Hart *et al.* [57] carried out detailed electrochemical studies to optimize the LCEC conditions for the determination of endogenous phylloquinone levels in human plasma. The detector consisted of a thin-layer cell containing a glassy carbon working electrode, which was operated in the reductive mode. Vitamin K_1 was separated from 10 cm^3 of plasma by solvent extraction with ethanol–hexane. The plasma extract was further purified by solid-phase extraction and normal-phase HPLC. LCEC was then carried out on a reversed-phase (C-8) column with a mobile phase containing methanol–acetate buffer pH 3.0 (95:5); the applied potential was −1.0 V (versus Ag/AgCl). The detection limit was found to be about 500 pg which was

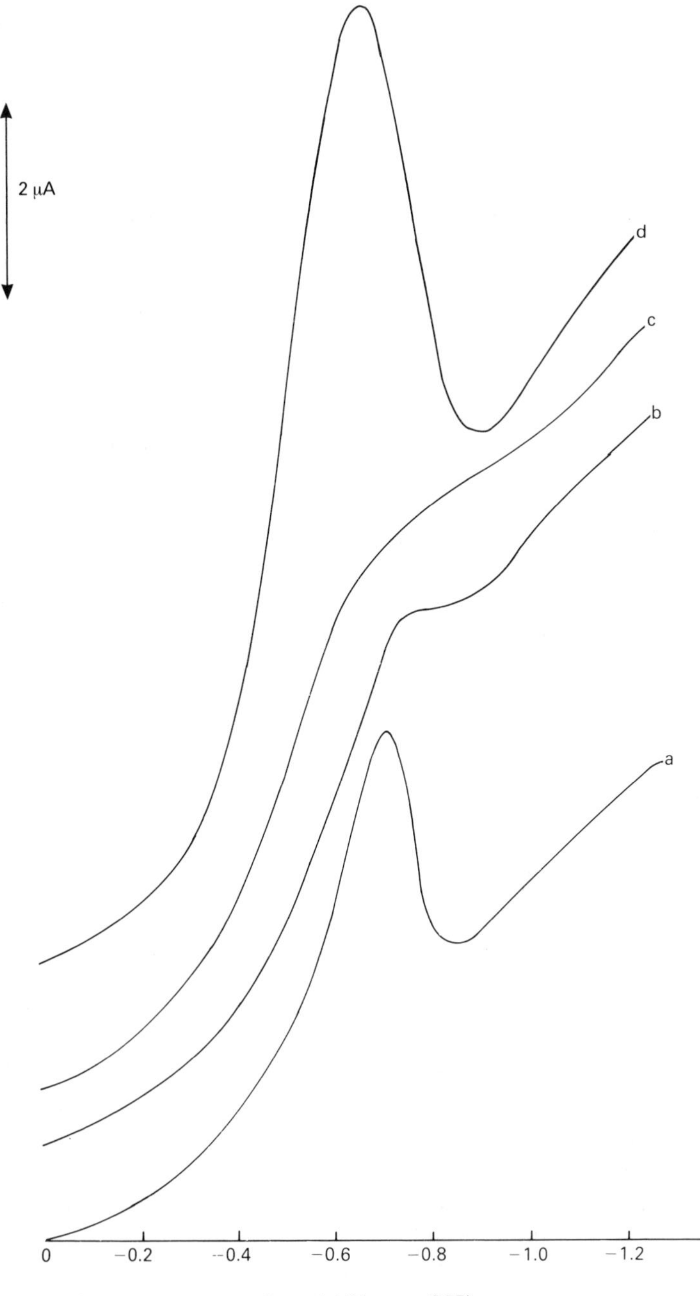

Fig. 4.15 — Linear-sweep voltammograms for vitamin K_1, after preconcentration, using different types of graphite to fabricate the electrode: (a) UFP graphite; (b) natural graphite; (c) synthetic graphite; (d) USP graphite. Percentage Nujol in each electrode was 25% w/w; electrolyte concentration was 50% ethanol–0.05-M acetate pH 5.0 and vitamin K_1 was 10^{-5} M. The accumulation time was 2 min with stirring and 30 s quiescence. (Reproduced from [70] by permission of the copyright holders, Royal Society of Chemistry.)

suitable for the determination of normal circulating levels (range, 0.08 to 1.24 ng cm^{-3}) following the preconcentration step.

Ueno and Suttie [72] have also investigated reductive-mode LCEC for the determination of phylloquinine in human serum. The method developed by these workers was different to that outlined above [57], but it was also suitable for the determination of normal circulating levels.

As mentioned earlier, vitamin K_3 can undergo reduction at carbon electrodes and an investigation has been carried out to compare the performances of reductive mode LCEC with LCUV [73]. In this study, a differential amplifier was employed [74] equipped with a cell containing two compartments. In one compartment a vitreous carbon electrode served as the working electrode for measurement of the background plus analyte; in the other compartment a vitreous carbon electrode measured the background current only. This mode eliminated the need for dearation of the mobile phase. The calibration graph for LCEC was linear from 200 ng to less than 0.5 ng; for the UV detector the lower limit of the linear range was about 8 ng and the upper limit was greater than 200 ng. Therefore, it was concluded that the differential amperometric detector was superior in terms of sensitivity.

In a further investigation by Hart *et al.* [75] it was desirable to reduce the volume of plasma required for the vitamin K_1 assay and to also readily measure depressed circulating levels of this vitamin; therefore, it was necessary to improve the sensitivity of the previously described method [57]. This was achieved by using the original HPLC system [57] and replacing the thin-layer cell with one containing two porous graphite electrodes in series [75]. These electrodes were operated in the redox mode: the upstream electrode (generator) was held at -1.3 V to reduce the quinone to the hydroquinone, which was detected downstream at the second electrode, which was held at 0 V, where re-oxidation to the quinone occurred.

There are several advantages of using the redox mode: first there is no need to deaerate the mobile phase continuously because oxygen products do not interfere at the detector electrode, and secondly, the background currents obtained at 0 V are much lower than those at the high negative potentials required for reduction; this results in better signal-to-noise ratios. In addition, the surface area of the porous graphite electrodes is greater than that of the glassy carbon electrode used in the earlier method [57]; therefore, higher concentrations of phylloquinone are electrolysed at the former. The overall effect of using a cell containing dual porous graphite electrodes in series in the redox mode was to increase the sensitivity of the method by an order of magnitude. This method was also found to be about 30 times more sensitive than a method using UV detection after HPLC [76].

The LCEC method using redox mode detection was applied to the determination of vitamin K_1 in normal subjects and patients suffering from osteoporosis; Fig. 4.16 shows some typical chromatograms. Clearly, in both cases well-defined peaks were obtained; the injected volumes were equivalent to only 1.3 and 1.8 ml of plasma respectively.

This method has been successfully applied to the measurement of depressed circulating phylloquinone levels in osteoporotic patients with subcapital [77] and trochanteric [78] fractures of the femur, as well as spinal crush fractures [79, 80]. Recently, it has been applied to determinations of circulating levels in patients with traumatic fractures including femur, ankle and humerus [81]; the levels were

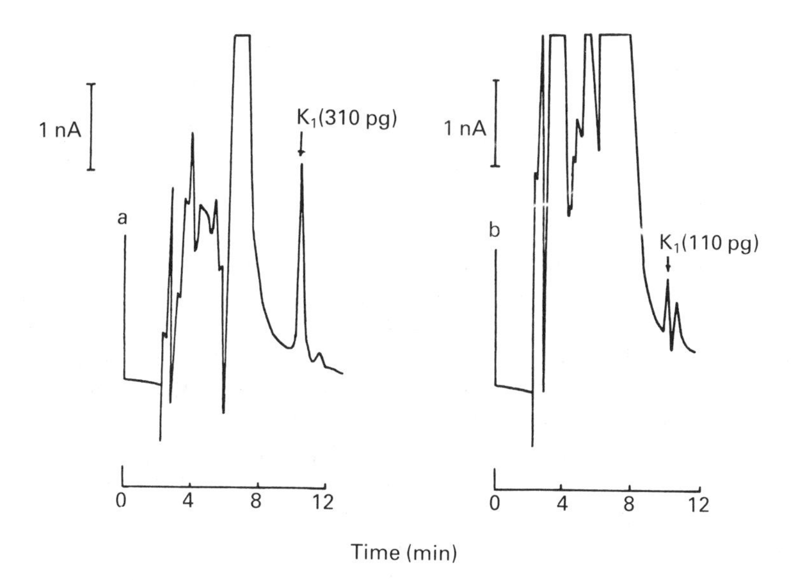

Fig. 4.16 — Chromatograms of plasma samples obtained by LCEC in the redox mode for (a) a normal subject (endogenous concentration of 240 pg cm^{-3} in plasma), (b) a patient with osteoporosis and fractured neck of femur (endogenous concentration of 60 pg cm^{-3} in plasma. (Reproduced from [75] by permission of the copyright holders, Royal Society of Chemistry.)

followed over various periods of time during fracture healing. This sensitive LCEC method should also be widely applicable in nutrition and clinical analysis, particularly where depressed vitamin K_1 levels are likely to be encountered.

This approach has also been employed in order to develop a method for the analysis of menaquinones [4]. A sample of plasma was taken from a patient suffering with hypertriglyceridaemia and subjected to the extraction procedure mentioned above; the resulting chromatogram showed the presence of menaquinones 6, 7, 8 as well as an elevated vitamin K_1 level in the plasma.

The series dual-electrode approach to vitamin K analysis has also been investigated by other workers. Haroon *et al.* [82] used coulometrically operated electrodes and successfully determined phylloquinone in rat liver; it was also shown that this could be used to determine a variety of menaquinones in a standard mixture. Langengberg *et al.* [83] compared the dual-electrode coulometric system with a system containing a coulometric generator electrode (upstream electrode) and an amperometric detector electrode (downstream electrode). The authors indicated that the second approach was slightly superior in terms of selectivity, but the detection limit was somewhat higher, i.e. 150 pg for the former and 280 pg for the latter.

A dual-series amperometric detection system has been investigated for the determination of phylloquinone and menaquinones in human and cow milk [84]. This detector was also operated in the redox mode and, therefore, the principles involved are as described above. Calibration graphs were linear in the range 2–50 ng and 1–39 ng for vitamin K_1 and MK-4 respectively; recoveries of added K vitamins were greater than 93%. Twenty-three samples of human milk were analysed by this method; the mean and standard deviation of phylloquinone was 21 ± 0.9 ng cm^{-3} and Mk-4 was 1.3 ± 1.0 ng cm^{-3}. In a few samples of human milk, small concentrations of MK-6 and MK-7 were found but MK-8 and MK-9 were not detected. Cow milk was analysed and concentrations were in the range 7.2–12.8 and 1.5–4.8 ng cm^{-3} for MK-4 and phylloquinone respectively; these results agreed with an HPLC method using fluorimetric detection.

A promising liquid chromatographic technique which combines electrochemistry and fluorimetry has recently been investigated by several research groups. The principle of operation is that the vitamin is electrochemically reduced and the product is measured by a fluometric detector. This approach has been applied to the determination of endogenous human plasma levels of vitamin K_1 [83] and vitamin K_1 epoxide [85], vitamin K_1 in raw and processed vegetables [86], and vitamins K_1 and K_2 in rat plasma [87]. It appears that electrofluorimetric detection can offer greater sensitivity and selectivity than UV detection [88], and is also simpler than fluorescence detection following post-column chemical reduction [89].

4.3 ELECTROANALYSIS OF WATER-SOLUBLE VITAMINS

4.3.1 Vitamin B_1 (thiamine)

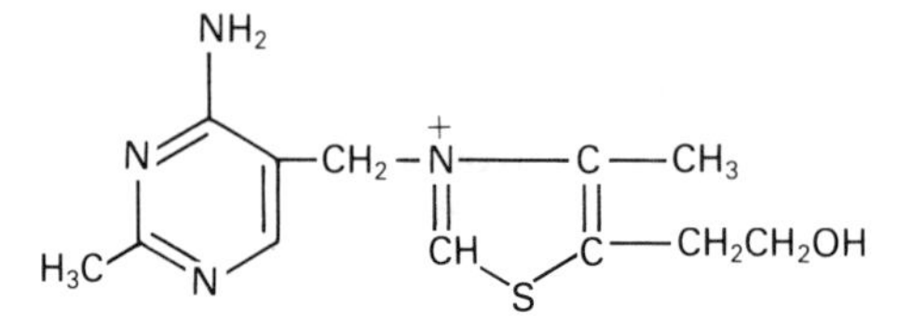

(XVIII)

Thiamine (XVIII) was discovered to be electroactive as long ago as 1941 when d.c. polarography was used together with a supporting electrolyte containing 0.1-M KCl [90]. Some years later, the same electrochemical technique was employed by Vergara *et al.* [91] for the determination of thiamine (T), its monophosphate (TMP) and pyrophosphate (TPP) esters in mixtures containing the three species. It was shown that by careful adjustment of buffer conditions d.c. polarography could be used to measure each of the compounds. In Britton–Robinson Buffer pH 9.0–9.3 only thiamine gave an anodic wave ($E_{1/2} = -0.4$ V versus SCE). This was the result of the oxidation of the thiol form of the vitamin at the DME as shown in (4.3).

$$RSH + Hg \rightarrow R\text{-}SHg + H^+ + e^- \qquad (4.3)$$

In pH 9.3–9.6, thiamine and its monophosphate gave a single anodic wave, but the thiazole ring of TPP was unreactive. However, in pH above 9.6 all three compounds were polarographically active and a single wave for all three compounds was obtained. The concentration of each component of the mixture could be determined by applying difference calculations. Calibration graphs were shown to be rectilinear over the range 7×10^{-5} to 1×10^{-3} M for T and TPP and 7×10^{-5} to 9×10^{-4} M for TMP.

A d.c. polarographic method was developed by Pedrero and Fonseca [92] that was based on a catalytic reduction process involving thiamine and Co(II). In this method, thiamine disulphide was first reduced to the thiol form which adsorbed onto the mercury electrode; this was followed by complexation of Co(II) with the adsorbed thiol. This complex produced a catalytic cobalt(II) pre-wave, thus regenerating the ligand, which was able to participate again in this cycle of reactions. The optimum supporting electrolyte for thiamine determination was found to contain 10^{-3}-M Co(II)–0.06-M borax–0.04-M boric acid (pH 7.85). In this medium, the relative error was found to be in the range +1.0 to −0.5% when concentrations of the vitamin between 1.0×10^{-6} M and 10×10^{-6} M were analysed.

Since DPP is inherently more sensitive than d.c. polarography it is perhaps not surprising that several research groups have investigated this technique for the determination of thiamine.

A method for the simultaneous determination of thiamine and thiourea in solutions for injections was reported by Kerchove and Bontemps [93]; this was based on the oxidation of the thiol moieties of the two species at the DME. Differential-pulse polarograms were recorded on solutions which had been treated with sodium hydroxide and peaks were exhibited at potentials of −0.42 V and −0.36 V for thiamine and thiourea respectively. This method was applied to injection solutions containing 0.2 mg of thiourea and 10 mg thiamine in a 5-cm^3 volume. Moorthy *et al.* [94,95] carried out detailed DPP studies to ascertain the nature of the reduction process for thiamine. These authors also showed that the addition of 0.04% Triton X-100 to solutions of 10 μM thiamine in 0.1-M acetate buffer pH 6.5 produced well-defined peaks (E_p − 1.35 V); this behaviour was retained for over three orders of thiamine concentration.

4.3.2 Vitamin B_2 (riboflavin)

The polarographic behaviour of riboflavin (XIX) has been reported by Lindquist and Farroha [65], who indicated that the molecule undergoes a reversible reduction process involving two electrons and two protons. In the same report, it was shown that the vitamin could be determined in some multivitamin preparations by DPP after simply diluting with phosphate buffer pH 7.2 and water. Detection limits were of the order 0.1 $\mu g\ cm^{-3}$.

DPP has also been applied successfully for the direct determination of riboflavin in the presence of the surfactants found in many pharmaceutical products [96]. In fact, it was shown that the magnitude of the peak current obtained in phosphate buffer increased by 100% when 200 $\mu g\ cm^{-3}$ of methocel (hydroxypropylmethylcel-

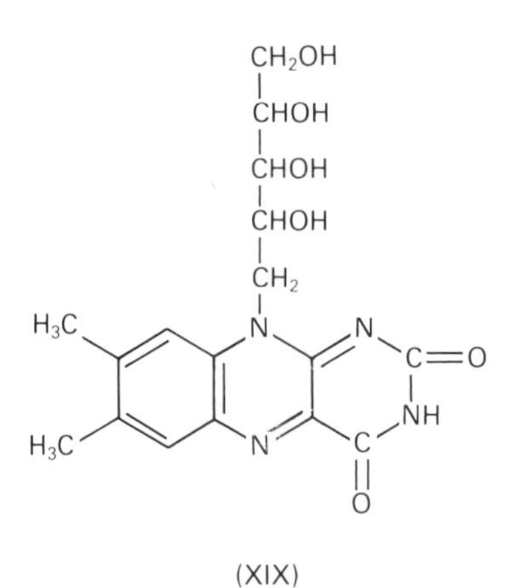

(XIX)

lulose) was added. In addition, the width of the peak at half height decreased from 100 to 80 mV, which indicated that the reduction process became more reversible; therefore, it appeared that this reduction process was coupled with adsorption behaviour.

This adsorption behaviour has recently been used very effectively to allow subnanomolar levels of riboflavin to be measured.

In one method [97–99], AdSV was carried out with a static mercury electrode. The preconcentration step involved spontaneous adsorption at a potential of -0.2 V; this was followed by the stripping step which employed a differential pulse waveform in a negative scan direction. The most suitable supporting electrolyte was found to be 0.001-M NaOH (Fig. 4.17); the limit of detection was shown to be 2.5×10^{-11} M. A similar AdSV method for riboflavin was developed by Grigorev *et al.* [100]; however, in this case an a.c. waveform was used in the stripping step. For concentrations between 0.2 and 25 nM the height of the peak at -0.65 V was recorded using an ammoniacal buffer, pH 11.75 (0.1 M); for concentrations between 0.2 and 1.5 μM the peak at -1.3 V was recorded with 1-M KCl as supporting electrolyte. A further AdSV study was reported by Sawamoto [101] in which adsorption of riboflavin, or its reduced form leucoflavin, was exploited in the analysis.

Surprisingly few applications of voltammetry to real samples containing riboflavin have been reported; however, one example which illustrates the potential usefulness in food analysis was given by Colugnati [102]. In this case riboflavin was determined in powdered milks and infant formula using cyclic voltammetry. Extraction of the vitamin was effected by grinding with sodium salicylate followed by suspension in pH-7.5 Sorenson buffer; this was centrifuged, and voltammetric analysis was performed on the resulting supernatant. It was shown that calibration plots were linear for 0–18 μg cm^{-3} of the vitamin.

4.3.3 Vitamin B_6

The vitamin B_6 group consists of six structurally related compounds: pyridoxal (PL, XX), pyridoxal 5′-phosphate (PLP, XXI), pyridoxine (PN, XXII) pyridoxine 5′-

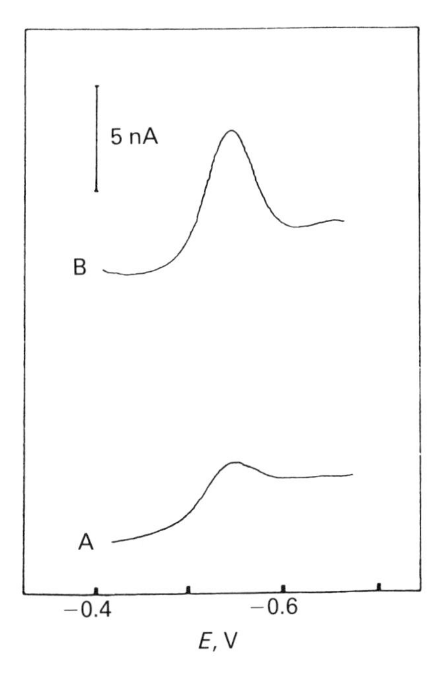

Fig. 4.17 — Differential pulse voltammograms for 5×10^{-10} M riboflavin. Preconcentration for 5 (A) and 30 (B) min at −0.2 V with 400 rpm stirring. 0.001-M NaOH solution. Scan rate, 5 mV s^{-1}; amplitude, 50 mV. (Reproduced from [98] by permission of the copyright holders, International Laboratory.)

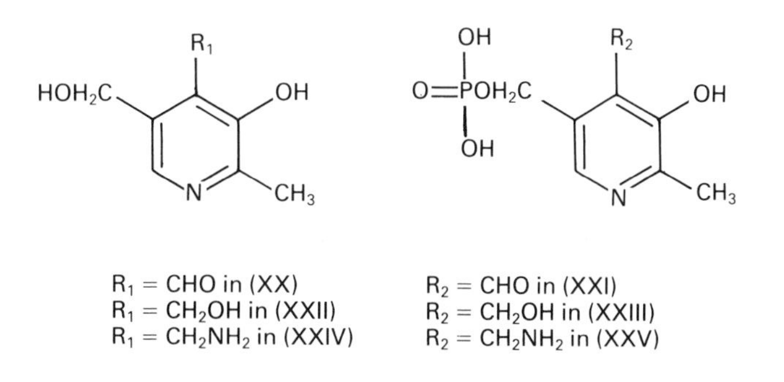

phosphate (PNP, XXIII), pyridoxamine (PM, XXIV) and pyridoxamine 5′-phosphate (PMP, XXV). Pyridoxine is the form usually found in foods and in pharmaceutical vitamin preparations. However, the predominant form in human plasma is PLP, an essential coenzyme for many enzyme reactions. Pyridoxal circulates at lower concentrations but seems to be the precursor for PLP; it is also the form required for transport into red cells, Pyridoxal 5′-phosphate is the coenzyme for the carboxylase effecting the formation of γ-carboxyglutamyl residues essential for bone mineralization [103].

4.3.3.1 Methods involving polarographic and voltammetric techniques

The polarographic behaviour of PLP has been the subject of several investigations [104–108]. In one study [104], this vitamer was found to give one reduction wave at pH values less than 10; it was shown that this was a result of a two-electron irreversible diffusion-controlled process. in other studies, Llor [105, 106] also investigated the electrochemical reduction of PLP and concluded that reduction occurred at the aldehyde group.

More recently, Izquierdo *et al.* [107, 108] investigated the electrochemical behaviour of PLP with a view to clarifying biological behaviour. These workers used both d.c. and DPP to investigate the behaviour over the pH range 0–7; the effect of buffer ionic strength, ethanol concentration and reactant concentration were also studied. At pH above 5.5 the aldehyde group was reduced in two-electron, two-proton process to produce the corresponding alcohol. However, under more acidic conditions, the process involved two one-electron steps, with the formation of an intermediate radical that could also dimerize; the reaction appears to be complicated by adsorption phenomena. Bearing in mind that the adsorption phenomenon has already formed the basis of sensitive AdSV methods for other vitamins, it may also be possible to use it in a similar way for PLP analysis.

The d.c. polarographic behaviour of pyridoxal has been investigated by Volke [109] and Manousek and Zuman [110, 111]. It was shown by these workers that the reduction process was kinetically controlled in acidic or neutral media. However, at pH values greater than 10 the reduction was considered to involve the addition of two electrons in a diffusion-controlled reaction; at these high pH values the limiting current had reached a maximum value. These results were later confirmed by Jacobsen and Tommelstad [112] who used phase-selective a.c. and differential-pulse polarography, in addition to d.c. polarography. It was also suggested [112] that the reaction was irreversible and that the reduction occurred in two steps; the first of these was observed as a well-defined DPP peak. Therefore, this peak was employed for the measurement of pyridoxal; in addition, it was shown that pyridoxine could be determined by DPP following its oxidation to pyridoxal with manganese dioxide. For the analysis of PN in multivitamin tables the supporting electrolyte was 0.1-M sodium hydroxide and the DPP peak appeared at a potential of -1.22 V versus Ag/AgCl. There was no interference from vitamins A, B_1, B_2, C, D, nicotinamide and pantothenate and excipients such as sugars, and magnesium stearate; the relative standard deviation, determined on 10 tablets containing 2 mg PN, was 3.1%.

In addition to reduction processes compounds of the B_6 group have also been found to undergo oxidation reactions at carbon electrodes; these have also been utilized for analytical purposes.

Hart and Hayler [113] reported on the cyclic voltammetric behaviour of PL and PLP at a planar glassy carbon electrode in an acetate buffer of pH 6.25 (Fig. 4.18). The oxidation reaction which causes peak I (Fig. 4.18b) for PL is irreversible; however, some reductive behaviour occurs on the reverse scan as seen by the

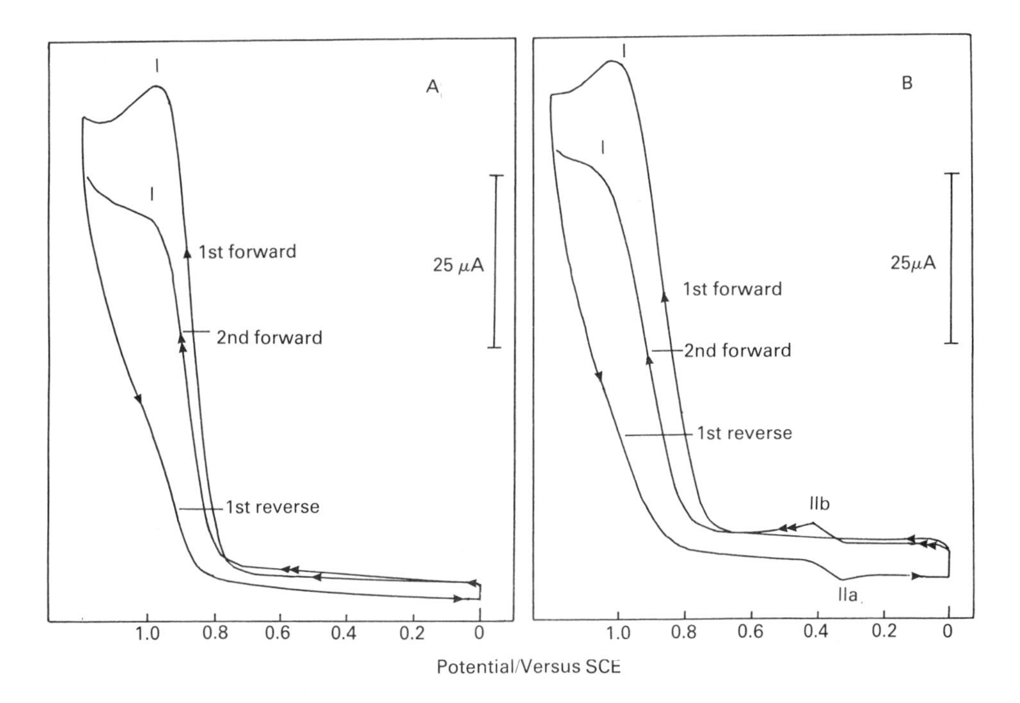

Fig. 4.18 — Cyclic voltammograms of (a) 0.0025-M PLP and (b) 0.0025-M PL. Supporting electrolyte, acetate buffer (pH 6.25); initial potential, 0 V; scan rate, 20 mV s^{-1}. (Reproduced from [113] by permission of the copyright holders, Royal Society of Chemistry.)

appearance of peak IIa. The process giving rise to this peak is quasi-reversible, as indicated by the occurrence of peak IIb on the second forward scan; the proposed mechanism of oxidation of PL is shown in Fig. 4.19. In contrast to PL, PLP shows only one oxidative peak (peak I, Fig. 4.18a) and no reductive peak was observed on the cyclic voltammograms after two successive scans; this peak was considered to be the result of a one-electron process to produce a free radical but the follow-up reactions were not elucidated. The oxidation reactions could be utilized in an LCEC method (described in the following section).

The oxidation reactions of B_6 vitamers have also been used in the analysis of pharmaceutical preparations. Linear-sweep voltammetry with a carbon paste electrode was employed for the determination of PL, PN and PM in a solution containing 0.25-M ammonia buffer pH 9.2 [114]. A linear calibration graph was obtained over the range 1×10^{-6} to 2×10^{-4} M. It was shown that PN could be determined directly in pharmaceutical products containing vitamins B_1, B_2, nicotinamide and calcium pantothenate. For the analysis of PN in preparations containing vitamin C and iron (II) it was necessary to introduce a chromatographic step. In contrast, a method involving DPV with a glassy carbon electrode was used for the direct measurement of PN even in the presence of iron (II) [115]. In this case, citrate buffer pH 4.0 was the supporting electrolyte and the anodic peak at +0.91 V versus SCE was used for quantitation.

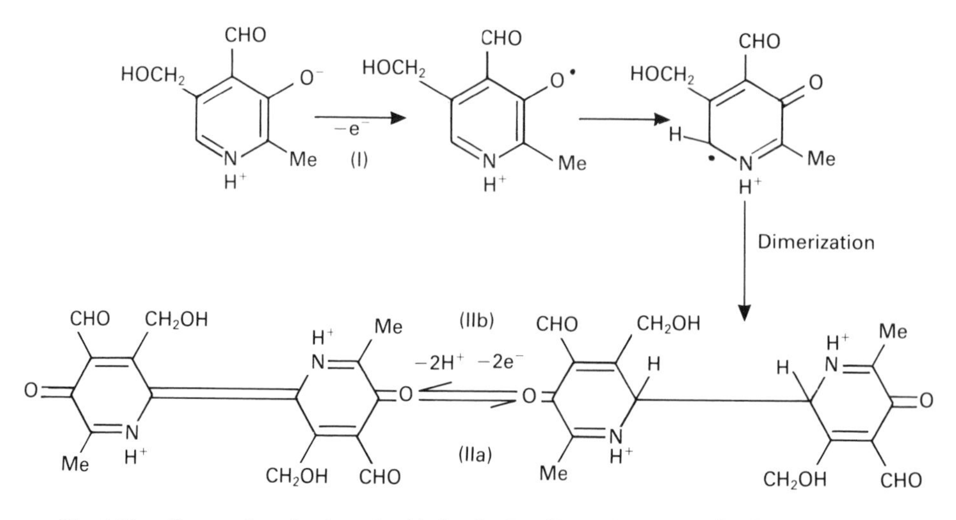

Fig. 4.19 — Proposed mechanism of oxidation for PL. (Reproduced from [113] by permission of the copyright holders, Royal Society of Chemistry.)

4.3.3.2 Methods involving LCEC

Wang and Hou [116] have determined PN in a pharmaceutical product using amperometric detection with a thin-layer cell containing a glassy carbon electrode set at +1.0 V; the mobile phase consisted of acetonitrile–0.1-M phosphate buffer pH 6.5 (6:94) and the stationary phase was ODS. Simple extraction of the vitamin was performed on one tablet; this involved grinding it to a powder, dilution with water and filtration. The retention time of PN in this system was about 5 min and there were no interfering peaks from other compounds present in the tablet.

The determination of pyridoxal phosphate in biological fluids presents some difficulty owing to the low concentrations expected as well as the possibility of interference from naturally occurring compounds; another difficulty is in separating PLP from the other five structurally related vitamers.

Allenmark *et al.* [117] developed an LCEC method for PLP which was based on the measurement of the dopamine produced by the reaction of L-DOPA with tyrosine apodecarboxylase; PLP acts as a coenzyme in this reaction and the amount of dopamine produced is dependent on the PLP concentration. The dopamine concentration was determined by amperometric detection with a thin-layer cell containing a carbon paste electrode, following separation on a cation-exchange column. The analysis could be performed on 100 μl of plasma and the sensitivity was adequate enough to measure 1 ng of PLP. It was suggested that this LCEC method was more convenient than radioenzymatic procedures.

A similar strategy was adopted by Lequeu *et al.* [118] for the determination of plasma PLP levels. This method involved PLP-dependent enzymatic decarboxylation of L-tyrosine to tyramine with tyrosine apodecarboxylase; the tyramine produced was measured and this was related to the PLP concentration. The LCEC system incorporated a reversed-phase column and a thin-layer amperometric cell equipped with a glassy carbon electrode; this was poised at +0.85 V versus Ag/AgCl. Calibration graphs were linear over the range 5–80 nM and the detection limit was

1.3 nM. The normal circulating plasma level found by this method was 65.1 nM±4.7 (n=30) which agreed with the data from other sources. The ease and simplicity of this procedure allowed more than 25 plasma samples to be assayed per working day.

Hart and Hayler [113] reported initial studies on the detection of plasma PL and PLP using coulometric dual-series electrodes in the screen mode. These early studies showed that it was possible to separate PL and PLP from PM, PMP, and PN using an ODS column; for plasma it was necessary to introduce a sample clean-up step, with a strong cation exchanger, having first removed the proteins. In addition, PLP needed to be hydrolysed to PL with acid phosphatase because it was not retained on the cation exchanger. Fig. 4.20 shows the chromatograms obtained for a normal subject

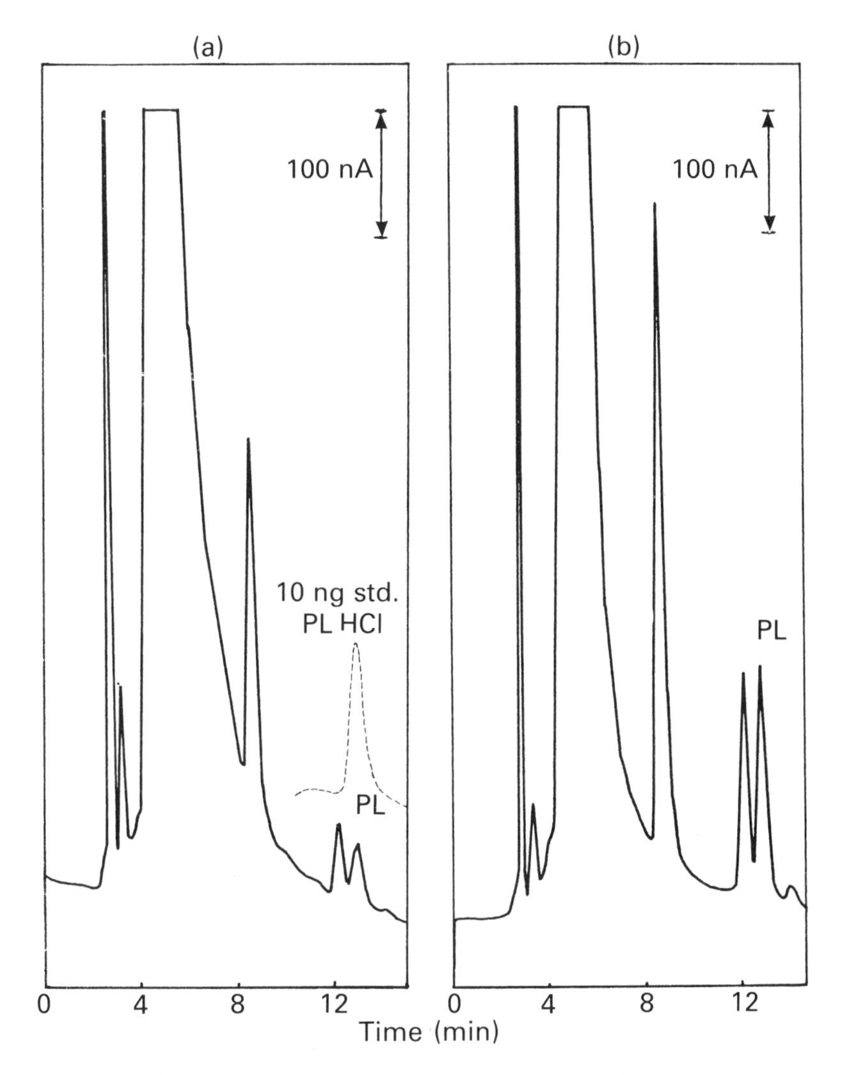

Fig. 4.20 — LCEC chromatograms of (a) normal plasma without acid phosphatase and (b) with acid phosphatase. Electrode 1, +0.6 V; electrode 2, 1.0 V; guard cell, 1.5 V. (Reproduced from [113] by permission of the copyright holders, Royal Society of Chemistry.)

before and after acid phosphatase treatment. While this clearly demonstrates potential for the analysis of plasma PL and PLP, some problems did occur which were related to inconsistent recoveries of PL from plasma during the method. However, if this problem can be resolved the LCEC procedure should be applicable to endogenous levels since the detection limits obtained on standards were 200 pg [3].

4.3.4 Folic acid

4.3.4.1 Methods involving polarography and voltammetry

Polarographic methods for the analysis of folic acid (pteroylglutamic acid, PteGlu:

(XXVI)

XXVI) are based on the cathodic signals produced as a result of reduction processes. Kwee [119] has described the mechanism of reduction and this is shown in Fig. 4.21.

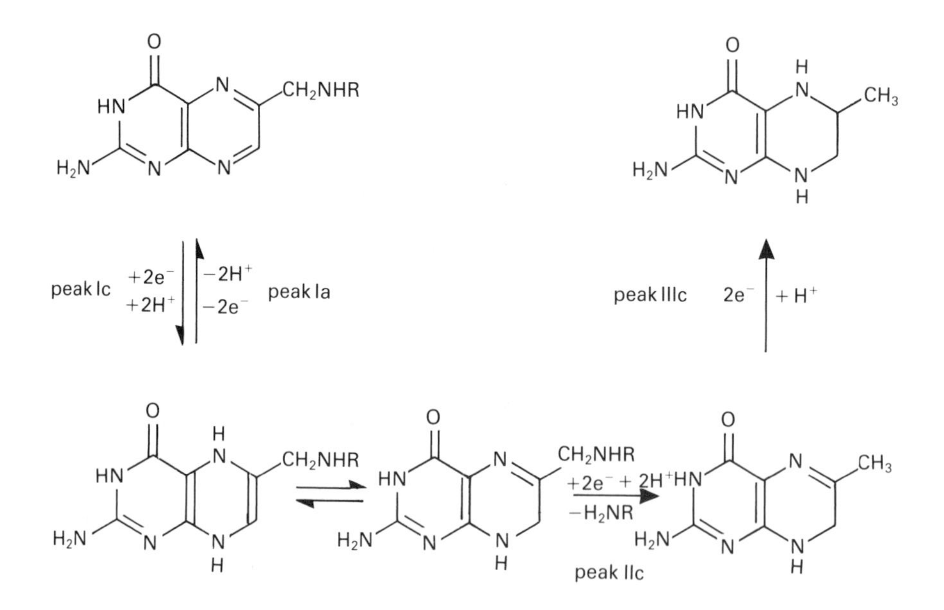

Fig. 4.21 — Proposed mechanism of reduction of folic acid. (Reproduced from [119] by permission of the copyright holders, Elsevier Science Publishers.)

Jacobsen and Bjornsen [120] also studied the electro-reduction of folic acid using a.c. and d.c. polarography and cyclic voltammetry. Fig. 4.22 shows the cyclic

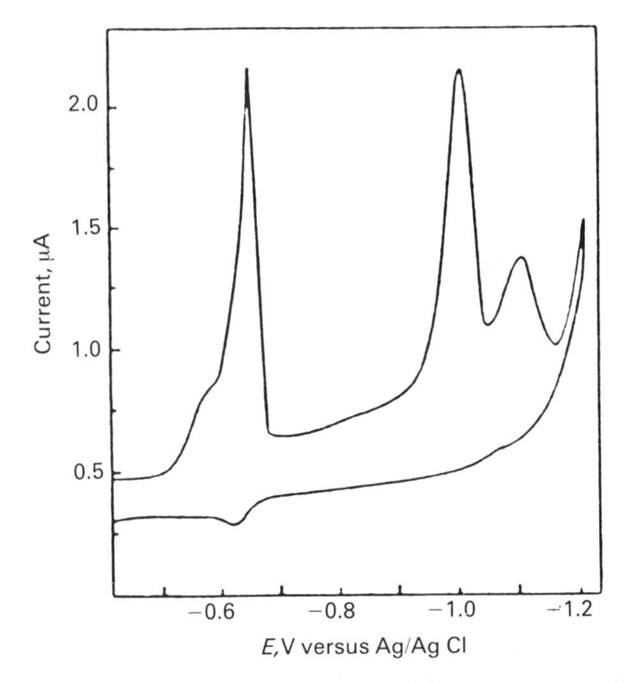

Fig. 4.22 — Cyclic voltammogram of 0.02-mM folic acid in 0.1-M acetate buffer pH 5.5. Scan rate 0.1 V s^{-1}. (Reproduced from [120] by permission of the copyright holders, Elsevier Science Publishers Physical Sciences and Engineering Division.)

voltammogram of folic acid at pH 5.0, and clearly this behaviour agrees with the mechanism proposed by Kwee [119].

In addition, the cyclic voltammetric behaviour of folic acid indicated that the reduction of the vitamin was coupled with adsorption characteristics. For analytical determinations. Jacobsen and Bjornson [120] found that phase-sensitive a.c. polarography was even superior to DPP by a factor of 10; consequently, the former was recommended for analysis. Using the a.c. technique, folic acid could be measured directly in pharmaceuticals containing large amounts of iron. Precipitation of hydrated iron oxide was prevented by the addition of diethylenetriamine pentaacetic acid. The limit of determination was 200 ng cm^{-3}, which was suitable for the analysis of one tablet containing 30 μg of the vitamin. A less sensitive d.c. polarographic method has also been described for the determination of folic acid in pharmaceutical products containing iron (II) [121]; however, this was still sensitive enough to measure the vitamin at 1.7 μg cm^{-3}.

The adsorption behaviour of folic acid reported by Jacobsen and Bjornson [120] has been further exploited in order to develop very sensitive methods of analysis.

Fernandez Alverez *et al.* [122] developed a method involving AdSV at a static mercury drop electrode; parameters including pH and ionic strength of buffers, buffer composition, as well as accumulation potential and time were investigated. For the analysis of human serum [123] the appropriate conditions were; supporting electrolyte, 0.1-M acetate buffer pH 5.0; preconcentration time 70 s at open circuit in quiescent solution. In addition, phase-sensitive a.c. voltammetry was found to be more sensitive than either DPV or LSV for the measurement of adsorbed analyte. The appropriate conditions were: frequency 75 Hz; scan rate 10 mV s^{-1}; potential range −0.35 to −0.70 V; the potential of the a.c. voltammetric peak was −0.65 V. It was necessary to carry out a preliminary separation of the vitamin from serum using solid-phase extraction on a C-18 cartridge. The detection limit was 5×10^{-9} M, with an overall precision of 9.9% at 10^{-7} M (n=7); the recovery was 57%.

Luo [124] described the AdSV behaviour of folic acid at an HMDE and its application to the determination of the vitamin in pharmaceutical tablets and urine samples; in this case the differential pulse waveform was used in the measurement step and a detection limit of about 1×10^{-10} M was obtained. In this method, the optimum supporting electrolyte was found to be 0.1 M sulphuric acid; the preconcentration time was usually 2 min at a potential of −0.3 V versus Ag/AgCl.

Another example of the application of AdSV for folic acid analysis was reported by Maali *et al.* [125]; in this case, square wave voltammetry was employed in the stripping step. The conditions recommended in this method were: preconcentration at an HMDE at −0.2 V versus Ag/AgCl; scan rate 200 mV s^{-1}; frequency 100 Hz and pulse amplitude 20 mV; the supporting electrolyte consisted of 0.03-M $NaClO_4$. The detection limit was found to be 2×10^{-11} with a 5-min deposition time; therefore, this would appear to be the most sensitive of the AdSV methods reported here.

Folic acid has also been shown to undergo oxidation at a glassy carbon electrode; the oxidation reaction was considered to occur at the 4-hydroxy moiety [126]. A well-defined DPV signal occurred at a potential of +0.63 V when the supporting electrolyte consisted of 3% m/v dibasic phosphate buffer. A simple method was described for the extraction of the vitamin from a multivitamin preparation containing vitamins B_1, B_2, C, nicotinamide and ferrous fumarate; these did not interfere in the voltammetric determination of folic acid.

4.3.4.2 Methods involving LCEC

The naturally occurring folic acid derivatives function as essential coenzymes in the biosynthesis of protein and nucleic acids [127]. In man, 5-methyltetrahydrofolic acid (5-MeTHF,5-methyl-H_4PteGlu) is the main analogue found in plasma [128,129]; it ranges from 3 to 16 ng cm^{-3} in normal human serum [130].

An LCEC method for the determination of 5-MeTHF in plasma and spinal fluid was developed by Lankelma *et al.* [131]. In order to separate the compound from plasma protein, precipitation with trichloroacetic acid was used; it was necessary to add ascorbic acid to the plasma during the procedure to prevent oxidation of 5-MeTHF. After centrifugation an aliquot of supernatant was injected onto a reversed-phase column for a combined concentration and first separation step; this was followed by an extra separation on an anion exchanger. For the concentration stage an eluent solution containing 0.015-M citrate in 0.015-M phosphate buffer (pH 4.95) was used; sodium azide was added to prevent the growth of microorganisms. The analytical eluent consisted of 0.05-M sodium phosphate buffer (pH 4.95)–methanol (4:1) and contained 0.001-M sodium chloride for the electrochemical detection. The detector contained a glassy carbon working electrode with a large surface area 92.65 cm^2 and the applied potential was set at +0.3 V versus Ag/AgCl. The method had a detection limit of 2×10^{-9} M (0.9 ng cm^{-3}) for both plasma and spinal fluid; the recoveries of 5-MeTHF from both of these samples was 100%.

This LCEC procedure was applied to patient plasma and spinal fluid samples and the results were compared with those obtained with a bioassay; some deviation was reported due to effects of interferences and standardization procedures for the bioassay.

A further example, which illustrates the potential of LCEC to separate and detect mixtures of folates, was recently described by Kohashi *et al.* [132]. A standard mixture containing PteGlu, tetrahydrofolic acid (H_4PteGlu), 5-Methyl-H_4PteGlu, 5-formyl-H_4PteGlu, and 10-formyl-H_4PteGlu was readily separated on a 5-μm phenyl-bonded-phase column; a glassy carbon working electrode (set at +0.75 V or +0.35 V) was employed in the detection system. The mobile phase consisted of 0.05-M potassium dihydrogen phosphate (pH 3.5) containing 0.1-mM disodium EDTA and 15% methanol. The method was applied to serum extracts from a human subject and a rat; Fig. 4.23 shows that the peak corresponding to 5-MeTHF was

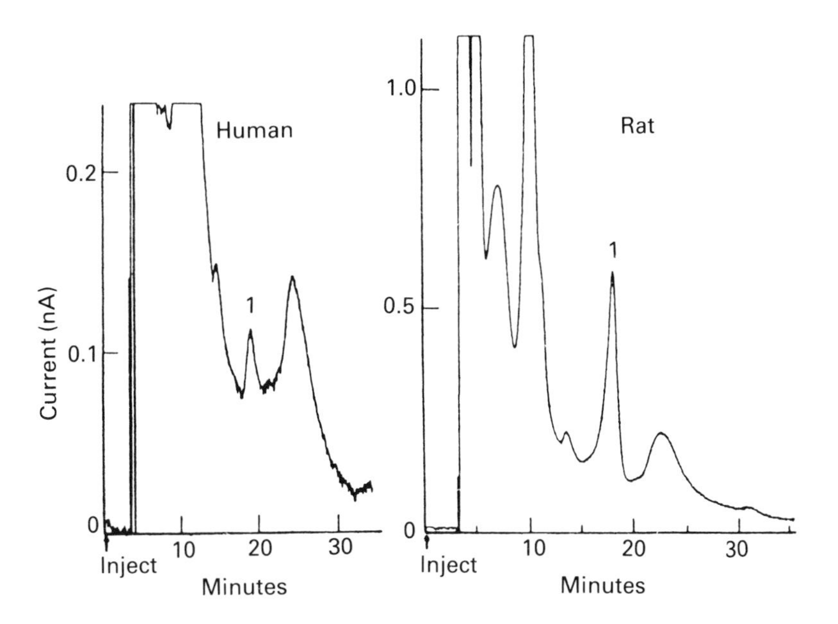

Fig. 4.23 — LCEC chromatograms of serum extracts form a human subject and a rat. Detector potential was set at +0.35 V versus Ag/AgCl. Peak 1 = 5-methyl-H_4PteGlu. (Reproduced from [132] by permission of the copyrignt holders, Elsevier Science Publishers Physical, Sciences and Engineering Division.)

well-resolved from other constituents of the samples. Serum levels of 5-MeTHF, before eating, in human and rat were 2.8 ± 1.2 ng cm^{-3} (n=5) and 10.0 ± 3.1 ng cm^{-3} (n=5), respectively.

4.3.5 Nicotinamide: methods involving polarographic and voltammetric techniques

A detailed voltammetric study was carried out by Jacobsen and Thorgersen in order to elucidate the mechanism of reduction of nicotinamide (XXVII) [133]. A single well-defined cathodic signal was obtained with 2-M sulphuric acid and with 0.1-M sodium hydroxide. In both media, nicotinamide was found to undergo a diffusion-

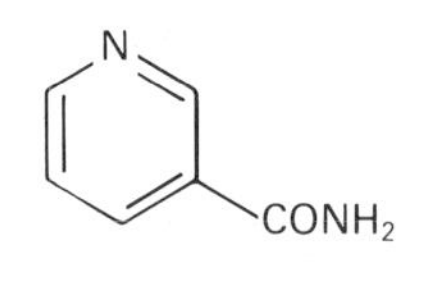

(XXVII)

controlled, irreversible, two-electron reduction; three protons were consumed in acidic solution and two protons in alkaline solutions. The proposed mechanisms of reduction, under these conditions, are illustrated in Fig. 4.24.

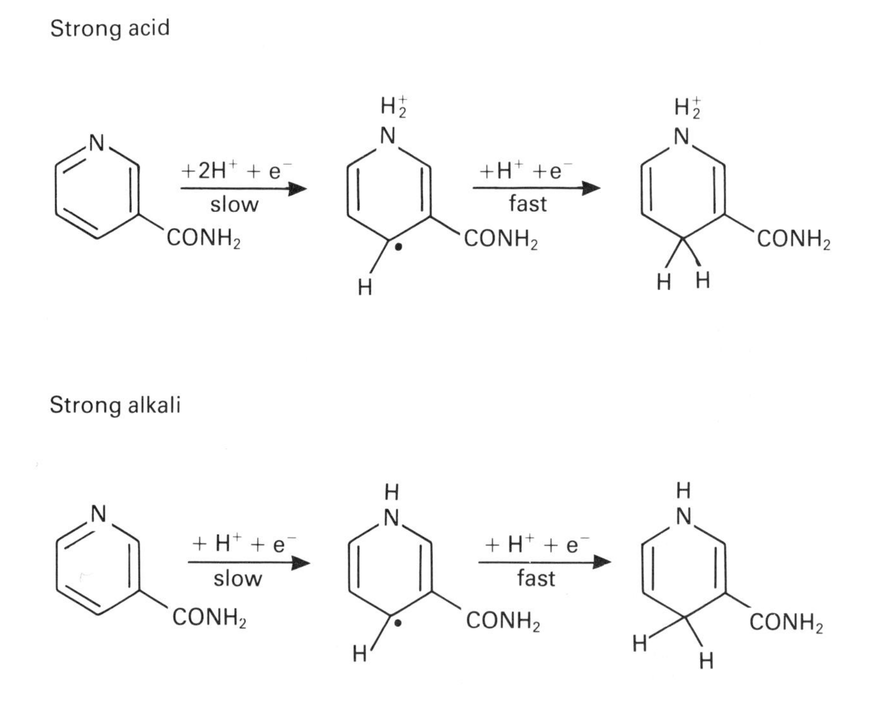

Fig. 4.24 — Proposed mechanism for the reduction of nicotinamide.

The same group [112, 133] have shown that DPP could be used in a simple, rapid method for the determination of nicotinamide in multivitamin tablets; the DPP peak appeared at a potential of −1.7 V versus Ag/AgCl with a supporting electrolyte consisting of 0.1-M sodium hydroxide. In this method, there was no interference from other vitamins or excipients present in the preparations. The same electroanalytical technique was also used by Lindquist and Farroha [65] for the analysis of nicotinamide in complex pharmaceutical preparations. However, in this case 0.1-M LiOH was chosen as the supporting electrolyte because the reduction of lithium ions occurs at very negative potentials; calibration graphs were found to be linear over the concentration range 1.0–0.02 μg cm^{-3}. Another example of the polarographic determination of nicotinamide in pharmaceutical preparations was described by Moore [134]; however, this method was rather more complex and involved chromatographic separation and eluent evaporation steps.

4.3.6 Electroanalysis of other B vitamins

This section contains examples of electroanalytical methods for several B vitamins that have not been discussed in previous sections (see Table 4.1).

4.3.7 Vitamin C

4.3.7.1 Methods involving polarographic and voltammetric techniques

Vitamin C (L-ascorbic acid, (XXVIII; Fig. 4.25) has probably been the subject of more electrochemical studies than any of the other vitamins [1]. This vitamin readily undergoes oxidation at both mercury and carbon electrodes [139,140]; the process involves the reversible loss of two electrons and two protons to produce dehydroascorbic acid, which is followed by an irreversible solvation reaction (Fig. 4.25).

Polarographic methods have been developed to measure ascorbic acid in a variety of solid and liquid foods. For the analysis of fruit juices and lemonade [141] a relatively simple procedure was adopted: a 25 cm^3 aliquot of sample was mixed with 10 cm^3 of a supporting electrolyte consisting of 2-M acetate buffer with added oxalic acid; this solution was deaerated with nitrogen followed by oxidative DPP. Amin [142], described a similar DPP method for the determination of vitamin C in fruit juice which involved dilution with 8% acetic acid.

DPP has also been applied to the determination of ascorbic acid in vegetables and fruits. Lau [143] recommended extraction of these foods with a mixture of 1% oxalic acid/2% TCA/1% Na_2SO_4; simple filtration was used to remove residues and 2-M acetate buffer was employed to keep the pH at 4.5. The optimun DPP conditions were modulation amplitude 50 mV, scan rate 2 mV s^{-1}, and drop time 1 s. The precision was found to be 1.4% at the 1 μg cm^{-3} level; calibration graphs were linear in the range 0–0.02 μg cm^{-3}. In this method interferences were reported for Fe^{3+}, bromide, iodide and EDTA; the authors recommended quantitation by the method of standard addition, Kozar *et al.* [144] demonstrated that for some types of processed and fresh fruit, e.g. citrus fruits and currants, vitamin C could be readily determined by DPP; however, others, such as bananas and cherries, were unsuitable for analysis. In another study [145] it was shown that fruits and vegetables containing the potential polarographic interferent glutathione could be assayed for vitamin C provided that the level of this thiol was less than 50 μg cm^{-3} [145].

Polarography has been successfully utilized for the measurement of ascorbic acid in other types of solid foods. For example, in the analysis of potato pellet and biscuits, samples were first crushed and extraction of the vitamin was performed with acetate buffer pH 4.7 [146]. A peak was obtained at +0.15 V when using fast linear-sweep polarography; this peak was employed for quantitation by the method of standard addition. Another example of this type was described by Sabino and

Table 4.1 — Electroanalytical methods for the determination of various B vitamins

Vitamin	Electrochemical technique	Supporting electrolyte/ mobile phase	Comments	Ref
Biotin	LCEC with ODS column. Detector +1.4 V versus Ag/AgCl	Acetonitrile–0.05-M KH_2PO_4 (adjusted to pH 2.0) 15:85	Determination of biotin in multi-vitamin tablets. Powdered tablet in 50 cm^3 of 0.05 M KH_2PO_4 (pH 2), filter, inject 5 μl onto column. No interference B_1, B_2, B_6, C, niacinamide.	[135]
Vitamin B_{12a} (Hydroxocobalamin)	DPP	0.2-M Phosphate buffer pH 6.8	Two peaks at: (a) −0.12 V due to reduction $B_{12a} \rightarrow B_{12r}$ i.e. Co(III)→Co(II) (b) −0.92 V due to reduction $B_{12r} \rightarrow B_{12s}$ i.e. Co(II)→Co(I). Linear range 8×10^{-5} M to 10^{-3} M. Studies on mixtures containing B_{12a} and thiols.	[136]
Vitamin B_{12} (Cyanocogalamin)	Cathodic AdSV with HMDE	0.3-M Ammonium acetate	Preconcentration potential −1.45 V versus SCE; preconcentration time 5 min; scan rate 50 mV s^{-1}. Linear range ≃6 nM–100 nM; limit of detection 2 nM; Std. dev. 5.8% with 50 nM, Cu(II), Cd(II), Zn(II), Ni(II) or Co(II) did not interfere, but Pb(II) did.	[137]
Vitamin B_{13} (Orotic acid)	DPP	1.0-M Perchloric acid	*Analysis of urine* (*orotic aciduria*) 1.0 M $HClO_4$/1–8 cm^3 urine; DPP peak at −0.665 V versus Ag/AgCl. Linear range 1.54–15.4 μg cm^{-3}; detection limit 1.54 μg cm^{-3}. *Analysis of serum* 1.0 M $HClO_4$/1–3 cm^3 serum; filtered sample diluted to 20 cm^3. DPP peak at −0.665 V; linear range 1.54–15.4 μg cm^{-3}. Coeff. var. 1.8% for 15.4 μg cm^{-3}. Recovery of orotic acid 70%.	[138]

Zorzetto [147] who developed a method for vitamin C analysis of bread mixes. This method simply required homogenization of the sample in deionized water, followed by filtration and dilution with acetate buffer pH 4.6. Recoveries were found to be between 99 and 100%; emulsifiers did not affect the polarographic signal if the

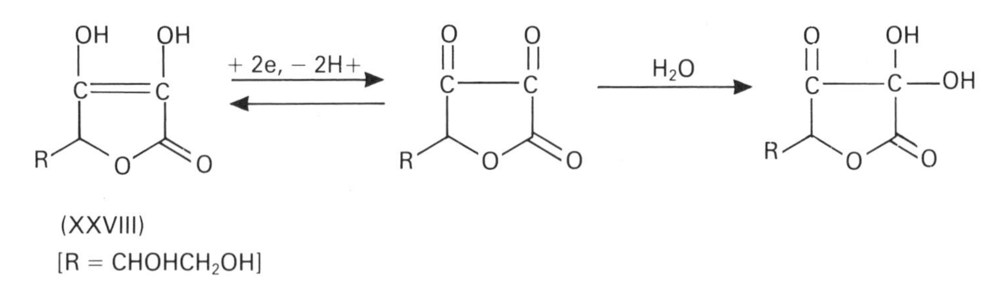

Fig. 4.25 — Mechanism for the oxidation of ascorbic acid.

concentration was less than 100 mg cm^{-3}. A DPP method was reported for the measurement of the vitamin C levels in wheat flour suspensions, but in this case 1.5% oxalic acid was recommended as the extraction medium [148]. Branca [149] has also recommended oxalic acid as an extraction medium for ascorbic acid from food samples. Using this same extraction solution, Gerhardt and Windmueller [150] were able to measure the vitamin C content of seasonings; it was possible to apply the method to cloudy or coloured solutions without further pre-treatment. For samples containing surface active agents a preliminary dilution step was used.

Voltammetric methods with carbon working electrodes have been readily applied to the determination of vitamin C in pharmaceutical preparations. Lindquist [151] has developed a method for the analysis of the vitamin in multivitamin tablets containing various amounts of iron (II), using LSV with a carbon paste electrode. In this method iron (II) was complexed by treating the sample with 5-nitro-1,10-phenanthroline in acetate buffer pH 3.7. The iron complex was oxidized at very positive potentials and did not interfere with the peak due to the oxidation of vitamin C (Fig. 4.26). In another application of this type [152] is was possible to determine

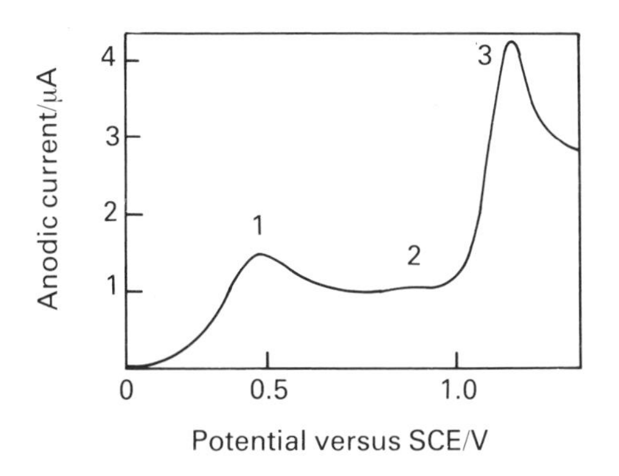

Fig. 4.26 — Voltammogram of an extract of a multivitamin tablet (Nutrivim) that contains vitamin C (100 mg) and iron (100 mg): 1, ascorbic acid; 2, pyridoxine; 3, iron complexed with 5-nitro-1,10-phenanthroline. (Reproduced from [151] by permission of the copyright holders, Royal Society of Chemistry.)

vitamin C in the presence of iron (II) without the addition of a complexing agent; this method involved DPV with a glassy carbon electrode and the supporting electrolyte consisted of citric acid pH 4.0.

A tubular carbon electrode was used by Mason *et al.* [153] to monitor vitamin C in multivitamin tablets. This method involved continuous-flow amperometry at a potential of +0.75 V versus SCE; the mobile phase consisted of 0.05-M acetic acid pH 3.0/0.01-M sodium nitrate. It was possible to analyse between 25 and 30 samples/h by the proposed procedure and it was suggested this might be amenable to automation.

Recently, it has been shown that chemically modified electrodes can be used to measure vitamin C at potentials lower than those required at conventional electrodes; it is possible that this approach could eliminate potential interferences. Wring and Hart [154] developed a method for the determination of vitamin C in single-vitamin and multivitamin preparations; the working electrode consisted of a carbon-epoxy composite electrode modified with the electron mediator cobalt phthalocyanine (CoPC; described in Chapter 3). A simple sample preparation procedure was employed that involved crushing a tablet and dissolving this in 50 cm^3 of 0.05-M phosphate buffer pH 5.0. An aliquot of the mixture was filtered and 100 μl of the filtrate was added to the electrochemical cell containing 20 cm^3 of fresh phosphate buffer pH 5.0. Amperometry was then performed in stirred solutions using a potential of +0.25 V versus SCE, and quantitation was performed by the method of standard addition. Fig. 4.27 shows the amperometric current response for a commercial preparation and the subsequent standard additions; the arrows indicate the point of addition and, as is apparent, the response time is quite rapid. The recovery of ascorbic acid added to the preparation was found to be 97.6%; the precision was 5.5% (n=6) for tablets containing a mean level of 50.995 mg/tablet.

A similar approach to vitamin C determinations has been proposed by Petersson [155] who employed a platinum electrode coated with a monolayer of covalently bound ferrocene (CpFeCp). The overpotential for ascorbic acid oxidation was decreased by 150 mV at pH 2.2 compared with a bare platinum electrode. The proposed electrocatalytic mechanism of oxidation of vitamin C (AH_2) is shown in Eqs. (4.4) and (4.5)

$$\text{-CpFeCp} \underset{}{\overset{\text{Electrode}}{\rightleftharpoons}} \text{CpFe}^{+}\text{Cp} + e^{-} \qquad (4.4)$$

$$\text{2-CpFe}^{+}\text{Cp} + \text{AH}_2 \longrightarrow 2\text{-CpFeCp} + \text{A} + 2\text{H}^{+} \qquad (4.5)$$

The ferrocene modified electrode was useful for the voltammetric determination of vitamin C in natural fruit juices. Other types of chemically modified electrodes have also been reported for the oxidation of ascorbic acid; these include monolayers of benzidine, polymerized vinylferrocene, alkylamine-siloxane polymers containing hexacyanoferrate (III), and polyvinylpyridine films containing hexachloroiridate (IV) [156–159].

An amperometric sensor for ascorbic acid has been developed by immobilizing

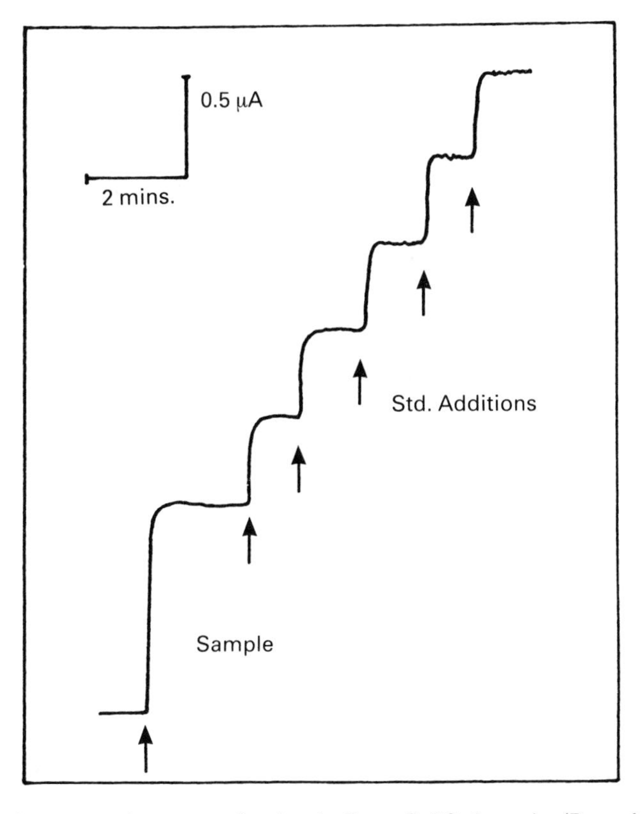

Fig. 4.27 — Amperometric response for vitamin C at a CoPC electrode. (Reproduced from [154] by permission of the copyright holders, Elsevier Science Publishers Physical Sciences and Engineering Division.)

ascorbate oxidase in the reconstituted collagen membrane and mounting the enzyme–collagen membrane on a Clark oxygen electrode [160]. The method is based on the amperometric measurement of the decrease in dissolved oxygen during the enzymatic reaction shown in (4.6).

$$\text{L-ascorbic acid} + \frac{1}{2}\,O_2 \xrightarrow{\text{Ascorbate oxidase}} \text{dehydroascorbic acid} + H_2O \qquad (4.6)$$

The response of the electrode was linear between 5×10^{-5} and 5×10^{-4}-M ascorbic acid; the precision was found to be better than 2.3% by 35 successive assays. The lifetime of the electrode was about 3 weeks. The method was applied to the determination of vitamin C in various fruits and the results compared well with those obtained by the 2,4-dinitrophenylhydrazine method.

Microelectrodes have been investigated for the *in vivo* measurement of vitamin C in the brain of the rat. In one study [161], a carbon fibre of radius 2.8 μm was employed as the working electrode. The normal pulse waveform was preferred because this minimized the adsorption of reaction products; this adsorption would have resulted in a depression of the signal. The endogenous level of vitamin C was calculated from the resulting *in vivo* voltammogram and a value of 500 μM was found. However, it should be added that the method does not differentiate beteeen vitamin C and dopamine and the values might not be reflecting the true vitamin C levels. The resolution between ascorbic acid and dopamine (or possibly 3,4-dihydroxyphenylacetic acid: DOPAC) was satisfactory when using a differential-pulse waveform at a carbon paste electrode [162]. In this *in vivo* investigation on the rat brain it was demonstrated that usually three peaks apeared on the voltammograms: the first at +0.13 V may be due to ascorbic acid, the second at +0.22 V may be due to dopamine and/or DOPAC, and the third at about +0.34 V might be an oxidation peak of 5-hydroxytryptamine and/or 5-hydroxyindole-3-acetic acid (Fig. 4.28)

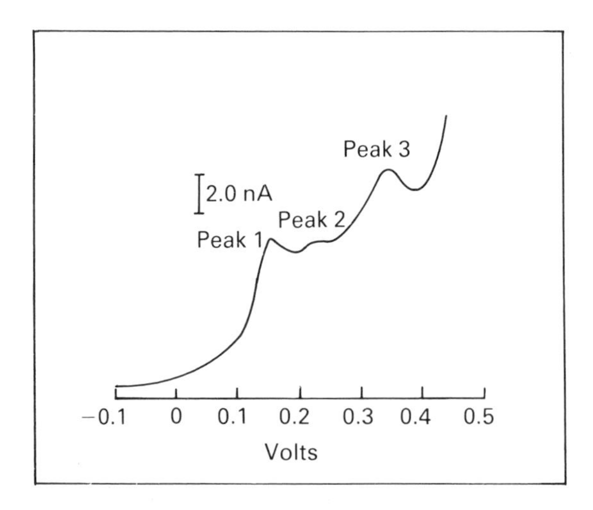

Fig. 4.28 — Differential-pulse voltammograms obtained *in vivo* from the striatum of a rat anaesthetized with chloral hydrate; note the three oxidation peaks at +0.13, +0.22 and 0.34 V. (Reproduced from [162] by permission of the copyright holders.)

Glassy carbon has been employed in many voltammetric methods for vitamin C analysis and recent investigations have shown that vacuum heat treatment may have beneficial effects by reducing the background charging current. Fagan *et al.* [163] used heat treatment at 725°C under high vacuum (less than 2×10^{-6} torr) which was shown to reduce the potential required for oxidation as well as increasing the voltammetric signal. Similar results were obtained by Deakin *et al.* [164] who also investigated the oxidation of vitamin C at heat-treated glassy carbon electrodes; it was concluded that surface activation did not result in electrocatalysis, but removed the impediments to electron transfer that exist at unactivated electrodes. Therefore, it would appear that this treatment of glassy carbon is worthwhile considering, particularly when applying these electrodes to vitamin C determinations in complex matrices.

4.3.7.2 Methods involving LCEC

Due to the inherent problem of interferences from biological samples, methods for the determination of vitamin C have been developed using liquid chromatography with electrochemical detection.

Reversed-phase columns have been investigated by Iryama *et al.* [165,166] for the analysis of body fluids. In this method, *m*-phosphoric acid/EDTA was employed to precipitate proteins from 0.5 cm^3 of serum and cerebrospinal fluid or 5 μl of urine. After filtration, a 10 μl aliquot of the filtrate was injected onto the column. The conditions and description of the chromatograms obtained on these samples were discussed in an earlier section (section 2.3.3), which dealt with the simultaneous determination of uric and ascorbic acids. A recent report by Behrens and Madere [167] described the possibility of using reversed-phase LCEC for the estimation of ascorbic acid and dehydroascorbic acid (DHAA) in tissues, biological fluids and foods. Dehydroascorbic acid was determined indirectly by converting it to ascorbic acid with homocysteine at pH 7.0–7.2 for 30 min at 25°C; it was necessary to treat one aliquot by this procedure and leave a second aliquot untreated, so that DHAA could be found by difference.

A slightly different approach to vitamin C analysis involved the use of reversed-phase LCEC with an ion-pairing agent added to the mobile phase; this was applied to the determination of vitamin C in human serum, plasma and leukocytes [168]. A more refined approach was described by Green and Perlman [169]; these workers used plasma ultrafiltration to remove protein prior to injection onto the reversed-phase column. The sensitivity of the method was improved and it was found that the column lifetime was increased.

A further method employing LCEC with an ion-pairing agent was reported by Tsao and Salimi [170]. In this report a number of ion-pairing agents were investigated for the purpose of separating ascorbic acid from isoascorbic acid, which were both present in a variety of biological samples; of the compounds investigated decylamine gave the best separation. The composition of the mobile phase was 40-mM acetate and 1-mM decylamine in methanol–water (15:85); the column was Ultrasphere ODS (5 μm) and the potential of the wax–graphite paste electrode was set at +0.7 V versus Ag/AgCl. This same research group have developed a similar method for the separation of ascorbic acid and ascorbic acid-2-sulphate that were present in biological matrices; however, in this case the ion-pairing agent was octylamine [171].

More recently, Kutnink *et al.* [172] described a paired-ion reversed-phase LCEC method that was capable of simultaneously determining ascorbic acid, isoascorbic acid and uric acid in human plasma. An ODS column was used together with a mobile phase containing 0.04-M sodium acetate/0.005-M tetrabutylammonium phosphate/0.2 mg cm^{-3} disodium EDTA, pH 5.25. The plasma samples were preserved with an equal volume of 10% metaphosphoric acid; these were then diluted by a factor of 10 with mobile phase and filtered through a 0.2 μm filter. Amperometric detection was carried out using a glassy carbon working electrode set at a potential of +0.6 V versus Ag/AgCl. Fig. 4.29 shows the LCEC chromatograms

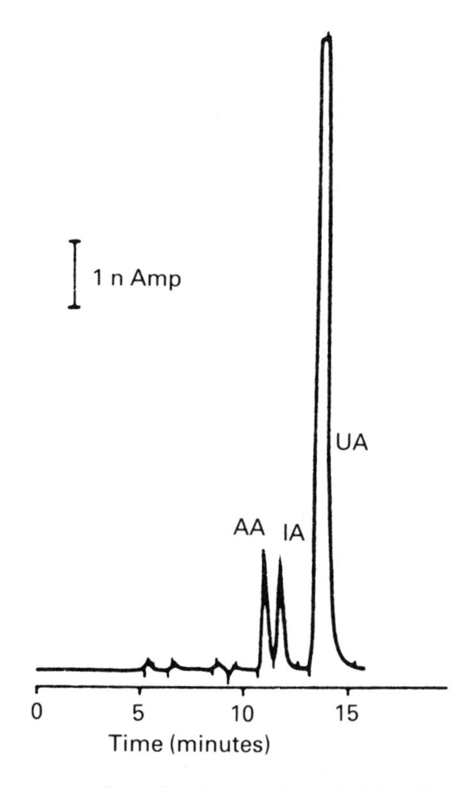

Fig. 4.29 — Chromatogram of unspiked human plasma (subject 1) containing 0.28, 0.27, and 2.49 mg dm^{-3} ascorbic acid, isoascorbic acid and uric acid, respectively. The plasma is preserved with an equal volume of 10% HPO_3 and diluted 10-fold with mobile phase prior to anlaysis. Injection volume is 10 μl. The detector–integrator response is within the 100-nA sensitivity setting for all three components. (Reproduced from [172] by permission of the copyright holders, Marcel Dekker.)

obtained for a normal subject and well-defined peaks are apparent for the three acids. The limit of detection for this method was found to be 0.25 ng for the three compounds using a sensitivity setting of 100 nA. It should be added that these authors also investigated two other ion-pairing agents, octylamine and decylamine; however, both proved to be unsatisfactory because they formed turbid complexes with HPO_3 when mobile phase was added to the preserved plasma samples.

In addition to reversed-phase chromatographic systems LCEC methods for vitamin C have been reported which involve separations with anion exchange columns. For the analysis of urine [173, 174], cold 50-mM perchloric acid was used to dilute the sample; analysis was performed with a Zipax SAX column using acetate buffer pH 4.75 as the mobile phase and the potential of the carbon paste electrode was set at 0.7 V versus Ag/AgCl. This system also provided well-resolved uric acid peaks.

Another LCEC method employing an anion exchange resin was described by Thrivikraman *et al.* [175]; the method was applied to the determination of vitamin C in mouse, rat and guinea pig brain samples. A further LCEC method using anion exchange resins was employed to separate vitamin C from the tissues of marine animals [176].

A cation exchange column has been investigated by Margolis and Davis [177] for the determination of vitamin C in human plasma; the mobile phase consisted of 0.1-M formic acid in aqueous 5% acetonitrile. Electrochemical detection was carried out with the detector set at +0.7 V and for comparison a UV detector was used at a wavelength of 268 nm. Fig. 4.30 shows the resulting chromatograms, from which it is

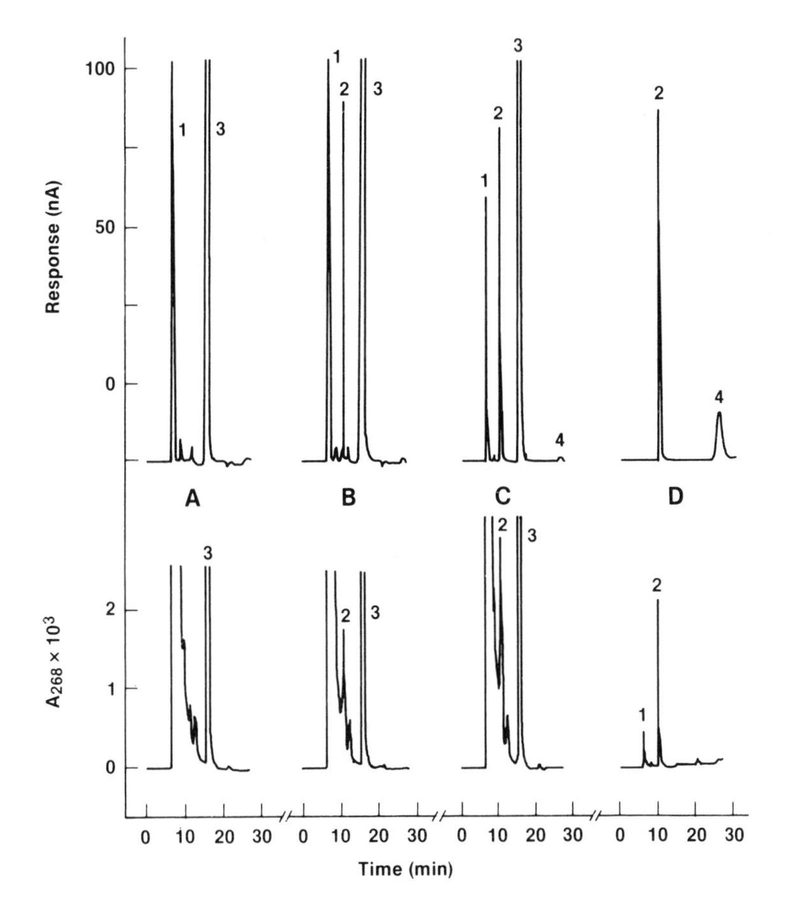

Fig. 4.30 — Separation of ascorbic acid (AA) in human plasma on an HPX87H Cation exchange column. (A) Unsupplemented plasma; (B) plasma with added AA, 4.80 mg dm^{-3}; (C) plasma of subject 1 (Table 4.2) assayed on day 52; (D) AA standard, 4.99 mg dm^{-3}, in acetonitrile/buffer (3:1 by vol.). Upper traces are electrochemical chromatograms; lower traces are ultraviolet chromatograms at 268 nm. Peak identity: 1, MPA; 2, AA; 3, uric acid; 4, DTT. (Reprinted with permission from *Clinical Chemistry* (1988), Vol. 34, Page 2217, Figure 1. Copyright American Association for Clinical Chemistry, Inc.)

clear that the sensitivity and selectivity were better when using the electrochemical detector; as is apparent the vitamin C peak was not well resolved from other UV peaks, whereas baseline resolution was observed with the LCEC system. The performance of the cation exchange column (method 1) was compared with a reversed-phase column (ODS) and a mobile phase containing an ion-pairing agent (sodium octyl sulphate) (method 2); as seen in Table 4.2 the results obtained by both methods were in good agreement.

Table 4.2 — Ascorbic acid concentrations in plasma of 11 normal subjects Age 18 to 40 years

Sex of subject	AA concentration, mg/l					
	Days stored[a]				Mean	SD[b]
	23	32	51	52		
F	7.8	7.3	7.8	6.9	7.5	0.4
F	19.9	19.2	18.7	18.9	19.2	0.5
F	8.7	8.4	8.6	8.5	8.6	0.1
M	10.8	10.8	10.9	10.8	10.8	0.0
F	12.1	11.8	12.3	12.0	12.1	0.2
F	7.6	8.1	7.7	8.2	7.9	0.3
M	8.7	8.8	8.5	8.6	8.7	0.1
M	9.8	10.0	9.5	10.4	9.9	0.3
M	15.1	15.9	14.9	15.4	15.3	0.4
M	7.9	7.9	7.6	8.0	7.9	0.1
F	16.3	16.5	16.3	16.0	16.3	0.2

[a]The samples on days 23 and 51 were analyzed by method 1 and the samples on days 32 and 52 were analyzed by method 2. Each value represents the mean of duplicate analyses.
[b]Standard deviation of a single measurement.

Several reports have indicated that the overpotential could be reduced for ascorbic acid oxidation at the glassy carbon detectors used in liquid chromatography. In one method [178] the electrode was electrochemically pretreated by holding it first at +1.75 V versus SCE for 5 min and then at −1.2 V for 10 s; this was done in a phosphate buffer of pH 4.7. Before treatment the oxidation potential was +0.6 V and after pretreatment it was +0.35 V. In another method [179] the glassy carbon electrode was modified by polishing the surface with 1 μm α-alumina particles on a deck of polishing cloth for 1 min, using a circular motion; the electrode was then rinsed with deionized water prior to LCEC.

A good review of analytical methods for the determination of ascorbic acid in biological samples, food products and pharmaceuticals was recently given by Pachla *et al.* [180]; this included a variety of electroanalytical methods for the vitamin.

4.4 CONCLUSIONS

In this chapter, examples have been described which illustrate the great potential of electroanalytical methods for determining vitamins in a wide variety of matrices.

In some cases it has been possible to directly measure vitamins by voltammetry/polarography in pharmaceutical preparation without interference from other vitamins, surfactants, or other excipients; coloured or turbid solutions have been analysed with no detrimental effects. This has resulted in rapid methods of analysis which are generally more convenient than spectroscopic means; the latter often require additional purification steps. It has also been posible to use voltammetric/polarographic techniques in direct methods for the analysis of certain types of foods; particularly fruits, fruit juices, vegetables and processed vegetables. In most of these methods the DPP waveform has been predominantly applied. However, square-wave voltammetry is now commercially available on most multi-mode polarographs; since it usually appears to be more sensitive and rapid than DPP it would seem worthwhile for workers to consider investigating this technique in future applications.

Electroanalytical procedures have also been developed for biological and biomedical investigations. For example, it has been possible to measure the clearance of vitamin K_1 from human plasma by DPP, following intravenous injection; in addition, *in vivo* voltammetry has been employed to measure endogenous levels of vitamin C in the rat brain.

Perhaps one of the most important advances in methodology for the determination of endogenous vitamin levels in biological fluids has been liquid chromatography with electrochemical detection (LCEC). This is a particularly difficult area of vitamin analysis because of the low levels expected (pg cm^{-3} concentrations in some cases) and the possibility of interference from other vitamins and naturally occurring substances. However, LCEC methods have now been developed for the measurement of normal circulating levels of the fat-soluble vitamins A, E and K and also depressed levels of A and K; it has been possible to monitor several vitamins together with their analogues/vitamers by this technique. LCEC methods also appear to be promising for monitoring some water-soluble vitamins in biological fluids; these include vitamin B_6, vitamin C and folic acid. No doubt work will continue in this area and it seems probable that eventually an LCEC method will be available for most of the vitamins.

Stripping voltammetry (involving adsorptive preconcentration) methods are particularly sensitive and a few examples have now appeared for vitamin analysis; however, few studies have examined complex samples by this technique. It may be possible to utilize this approach for complex matrices such as blood and urine by using a sample clean-up step with, for example, solid phase separations; these solid supports, constructed from a variety of materials, are commercially available in cartridge form. This would seem an area for future research.

Electrochemical sensors employing CMEs, which are based on electrocatalytic reactions, have been investigated for vitamin C; this has resulted in simple methods of analysis. A few examples have also appeared which illustrate the possibility of employing sensors based on coupling enzyme reactions with electrochemical devices; this type of approach should produce quite selective and yet simple procedures for vitamin analysis.

REFERENCES

[1] J. P. Hart, in *Investigative Microtechniques in Medicine and Biology*, J. Chayen and L. Bitensky (Eds.), Marcel Dekker, New York, 1984, p 199.
[2] J. P. Hart, *Trends. Anal. Chem.*, 1986, **5**, 20.

[3] J. P. Hart, in *Electrochemistry Sensors and Analysis*, M. R. Smyth and J. G. Vos (Eds), Elsevier Amsterdam, 1986, p 355.
[4] J. P. Hart, in *Vitamin K and Vitamin K-dependent Proteins: Analytical, Physiological and Clinical Aspects*, M. J. Shearer and M. J. Seghatchian (Eds) CRC Press, 1989, in press.
[5] H. A. Tyler, *Br. Nutr. Found. Nutr. Bull.*, 1986, **11**, 166
[6] R. Takahashi and I. Tachi, *Agr. Biol. Chem. Tokyo*, 1962, **26**, 771.
[7] E. J. Kuta, *Science*, 1964, **144**, 1130.
[8] S. M. Park, *J. Electrochem. Soc.*, 1978, **125**, 216.
[9] J. G. Osteryoung and R. A. Osteryoung, in *Electrochemistry Sensors and Analysis*, M. R. Smyth and J. G. Vos (Eds), Elsevier, Amsterdam, 1986, p 3.
[10] J. G. Osteryoung and R. A. Osteryoung, *Anal. Chem.*, 1985, **57**, 101A.
[11] C. Menicagli, S., Silvestri and L. Nucci, *Farmaco Ed. Prat.*, 1976, **31**, 244..
[12] S. A. Wring, J. P. Hart and D. W. Knight, *Analyst*, 1988, **113**, 1785.
[13] W. A. MacCrehan and E. Schonberher, *J. Chromatogr.*, 1987, **417**, 65.
[14] S. S. Atuma, J. Lindquist and K. Lunstrom, *Analyst*, 1974, **99**, 683.
[15] S. A. Wring and J. P. Hart, *J. Med. Lab. Sci.*, 1989, in press.
[16] J. P. Hart and P. H. Jordan, *Analyst*, 1989, **114,** 1633.
[17] D. W. Martin, Jr., in *Harpers Review of Biochemistry*, 19th Edition, D. W. Martin, Jr., P. A. Mayes and V. W. Rodwell, Lange, Los Altos, 1983, p 114.
[18] E. J. Kuta, *Science*, 1964, **144**, 1130.
[19] C. Menicagli, S. Silvestri and L. Nucci, *Farmaco Ed. Pr.*, 1976, **31**, 244.
[20] S. S. Atuma, K. Lundstrom and J. Lindquist, *Analyst*, 1975, **100**, 827.
[21] S. S. Atuma, *Trends in Anal. Chem.*, 1982, **1**, 339.
[22] J. Hernandez Mendez, A. Sanchez Perez, M. Delgado Zamarreno and M. L. Hernandez Garcia, *J. Pharm. Biomed Anal.*, 1988, **6**, 737.
[23] M. Delgado Zamarreno, A. Sanchez Perez, J. Hernandez Mendez and F. A. Martin Dominguez, *J. Pharm. Biomed Anal.*, 1989, in press.
[24] A. Sanchez Perez, M. Delgado Zamarreno, J. Hernandez Mendez and R. M. Sanchez Rodriguez, *Anal. Chim. Acta*, 1989, **225**, 247.
[25] I. Davidson, in *Polarography of Molecules of Biological Significance*, W. F. Smyth (Ed.), Academic Press, London, 1979, p 127.
[26] L. I. Smith, I. M. Kolthoff, S. Wawzonek and P. M. Ruoff, *J. Am. Chem. Soc.*, 1941, **63**, 1018.
[27] H. Schmandke, *Int. Z. Vitaminforsch*, 1965, **35**, 237.
[28] H. Smandke and H. Crohlke, *Clin. Chim. Acta*, 1965, **11**, 491.
[29] H. D. McBride and D. H. Evans, *Anal. Chem.*, 1973, **45**, 446.
[30] S. S. Atuma, J. Lindquist, *Analyst*, 1973, **98**, 886.
[31] J. Loeliger and F. Saucy, *Z. Lebensm. Unters. Forsch.*, 1980, **170**, 413.
[32] P. Deldime, B. Jacobsberg and M. Belhassine, *Anal. Lett.*, 1978, **A11**, 63.
[33] C. H. Brieskorn and K. Mahlmeister, *Z. Lebensm. Unters. Forsch.*, 1980, **171**, 348.
[34] K. Shiozaki, K. Fukui and T. Kitagawa, *Japan Analyst*, 1971, **20**, 438.
[35] J. Wang and B. A. Freiha, *Anal. Chim. Acta*, 1983, **154**, 87.
[36] M. D. Hawley, in *Laboratory Techniques in Electroanalytical Chemisty*, P. T. Kissinger and W. R. Heineman (Eds), Marcel Dekker, New York, 1984, p 463.
[37] J. H. Sung, S. H. Park, A. R. Mastri and W. J. Warwick, J. *Neuropathol. Exp. Neurol.*, 1980, **39**, 584.
[38] J. L. Rosenblum, J. B. Keating, A. L. Prensky and J. S. Nelson, N. *Engl. J. Med.*, 1981, **304**, 503.
[39] P. M. Suter and R. M. Russell, *Am. J. Clin. Nutr.*, 1987, **45**, 501.
[40] *Laboratory News*, Feb., 1989, p. 8.
[41] P. P. Chou, P. K. Jaynes and J. L. Bailey, *Clin. Chem.*, 1985, **31**, 880.
[42] P. P. Chou, P. K. Jaynes and J. L. Bailey, *Clin. Chem.*, 1985, **31**, 1027.
[43] M. Vandewoude, M. Clayes and I. De Leeuw, *J. Liq. Chromatogr.*, 1984, **311**, 176.
[44] M. C. Castle and W. J. Cooke, *Therap. Drug Monitor.*, 1985, **7**, 364.
[45] W. A. MacCrehan and E. Schonberger, *Clin. Chem.*, 1987, **33**, 1585.
[46] J. K. Lang and L. Packer, *J. Chromatogr.*, 1987, **385**, 109.
[47] J. K. Lang, K, Gohil and L. Packer, *Anal. Biochem.*, 1986, **157**, 106.
[48] G. W. Burton, A. Webb and K. A. Ingold, *Lipids*, 1985, **20**, 29.
[49] G. A. Pascoe, C. T. Duda and D. J. Reed, *J. Chromatogr.*, 1987, **414**, 440.
[50] T. Veda and O. Igarishi, *J. Micronutr. Anal.*, 1985, **1**, 31.
[51] G. W. Schieffer, *Anal. Chem.*, 1985, **57**, 2745.
[52] J. C. Vire and G. J. Patriarche, *Analysis*, 1978, **6**, 395.
[53] G. J. Patriarche, and J. C. Vire, in *Electroanalysis in Hygiene, Environmental, Clinical and Pharmaceutical Chemistry*, W. F. Smyth, (Ed.), Elsevier, Amsterdam, 1980, 209.

[54] J. P. Hart, and A. Catterall, *Anal. Chim. Acta*, 1981, **128**, 245.
[55] K. Takamura, and Y. Hayakawa, *J. Electroanal. Chem.*, 1974, **49**, 133.
[56] G. J. Patriarche, and J. J. Lingane, *Anal. Chim. Acta*, 1970, **49**, 241.
[57] J. P. Hart, M. J. Shearer, P. T. McCarthy and S. Rahim, *Analyst*, 1984, **109**, 477.
[58] O. S. Ksenzhek, S. A. Petrova, M. V. Kolodyazhnyi and S. V. Oleinik, *Bioelectrochem. Bioenerg.*, 1977, **4**, 335.
[59] G. Cauquis, and G. Marbach, *Experientia Suppl.*, 1971, **18**, 205.
[60] J. C. Vire, and G. J. Patriarche, *Analysis*, 1979, **7**, 144.
[61] J. C. Vire, G. J. Patriarche and G. D. Christian, *Anal. Chem.*, 1979, **51**, 752.
[62] K. Takamura and F. Watanabe, *Anal. Biochem.*, 1976, **74**, 512.
[63] J. P. Hart and A. Catterall, in *Electroanalysis in Hygiene, Environmental, Clinical and Pharmaceutical Chemistry, Analytical Chemistry Symposia Series*, Vol. 2, W. F. Smyth (Ed.), Elsevier, Amsterdam, 1980, 145.
[64] J. P. Hart, A. M. Nahir, J. Chayen and A. Catterall, *Anal. Chim. Acta*, 1982, **144**, 267.
[65] J. Lindquist and S. M. Farroha, *Analyst*, 1975, **100**, 377.
[66] J. V. Scudi, and R. P. Buhs, *J. Biol. Chem.*, 1942, **144**, 599.
[67] S. A. Akman, F. Kusu, K. Takamura, R. Chlebowski and J. Block, *Anal. Biochem.*, 1984, **14**, 488.
[68] J. C. Vire, V. Lopez, G. J. Patriarche and G. D. Christian, *Anal. Lett.*, 1988, **21**, 2217.
[69] J. C. Vire, N. Abo El Maali and G. J. Patriarche, *Talanta*, 1988, **35**, 997.
[70] J. P. Hart, S. A. Wring and I. C. Morgan, *Analyst*, 1989, **114**, 933.
[71] S. Ikenoya, K. Abe, T. Tsuda Y, Yamano, O. Hiroshima, M. Ohmae, and K. Kawabe, *Chem. Pharm. Bull.*, 1979, **27**, 1237.
[72] T. Ueno and J. W. Suttie, *Anal. Biochem.*, 1983, **133**, 62.
[73] K. Brunt, C. H. P. Bruins and D. A. Doornbos, *Anal. Chim. Acta*, 1981, **125**, 85.
[74] K. Brunt and C. H. P. Bruins, *J. Chromatogr.*, 1979, **172**, 37.
[75] J. P. Hart, M. J. Shearer, and P. T. McCarthy, *Analyst*, 1985, **100**, 1181.
[76] M. J. Shearer, P. Barkhan, S. Rahim, and L. Stimmler, *Lancet*, 1982 (ii), 460.
[77] J. P. Hart, A. Catterall, R. A. Dodds, L. Klenerman, M. J. Shearer, L. Bitensky and J. Chayen, *Lancet*, 1984 (ii), 283.
[78] L. Klenerman, B. Ferris and J. P. Hart, *J. Bone Joint Surg.*, 1988, **70B**, 286.
[79] J. P Hart, M. J. Shearer, L. Klenerman, A. Catterall, J. Reeve, P. N. Sambrook, R. A. Dodds, L. Bitensky and J. Chayen, *J. Endocrinol, Metab.*, 1985, **60**, 1268.
[80] J. P. Hart, R. Yaakub, M. J. Shearer, L. Klenerman, A. Catterall, J. Reeve, P. N. Sambrook, L. Bitensky and J. Chayen, *Clin. Sci.*, 1985, **68**, 29P.
[81] L. Bitensky, J. P. Hart, A. Catterall, S. J. Hodges, M. Pilkington, J. Chayen and A. Catterall, *J. Bone Joint Surg.*, in press.
[82] Y. Haroon, C. A. Schubert and P. V. Hauschka, *J. Chromatogr. Sci.*, 1984, **22**, 89.
[83] J. P. Langenberg and U. R. Tjaden, *J. Chromatogr.*, 1984, **305**, 61.
[84] H. Isshiki, Y. Suzuki, A. Yonekubo, H. Hasegawa and Y. Yamamoto, *J. Dairy Sci.*, 1988, **71**, 627.
[85] J. P. Langenberg and U. R. Tjaden, *J. Chromatogr.*, 1984, **289**, 377.
[86] J. P. Langenberg, U. R. Tjaden, E. M. De Vogel, and D. I. Langerak, *Acta Aliment*, 1986, **15**, 187.
[87] K. Kusube, K. Abe, O. Hiroshima, Y. Ishiguro, S. Ishikawa and H. Hoshida, *Chem. Pharm. Bull.*, 1984, **32**, 179.
[88] M. F. Lefevre, A. P. De Leenheer and A. E. Clayes, *J. Chromatogr.*, 1979, **186**, 749.
[89] K. Abe, O. Hiroshima, K. Ishibashi, M. Ohmae, K. Kawabe, and G. Katsui, *Yakugaku Zasshi*, 1979, **99**, 192.
[90] J. J. Lingane and O. L. Davies, *J. Biol. Chem.*, 1941, **137**, 567.
[91] T. Vergara, D. Martin and J. Vera, *Anal. Chim. Acta*, 1980, **120**, 347.
[92] P. S. Pedrero and J. M. L. Fonseca, *Analyst*, 1972, **97**, 81.
[93] C. Van Kerchove and R. Bontemps, *J. Pharm. Belg.*, 1982, **37**, 169.
[94] P. N. Moorthy, K. Kishore and K. N. Rao, *Proc. Indian Acad. Sci. Chem. Sci.*, 1981, **90**, 371.
[95] K. Kishore, P. N. Moorthy and K. N. Rao, *Indian J. Chem.*, 1979, **A17**, 206.
[96] E. Jacobsen, in *Electroanalysis in Hygiene, Environmental, Clinical and Pharmaceutical Chemistry*, W. F. Smeth (Ed.), Elsevier, Amsterdam, 1980, 227.
[97] J. Wang, D. Luo, P. A. M. Farias and J. S. Mahmoud, *Anal. Chem.*, 1985, **57**, 158.
[98] J. Wang, *Internatl. Lab.*, 1985, **15**, 68.
[99] J. Wang, *Internatl. Clin. Prods.*, 1986, **5**, 50.
[100] V. I. Grigorev, Y. F. Milyaev and L. N. Balyatinskaya, *Zh. Anal. Khim.*, 1985, **40**, 736.
[101] H. Sawamoto, *J. Electroanal. Chem.*, 1985, **186**, 257.
[102] L. Colugnati, *Boll. Chim. Ig., Parte Sci.*, 1985, **36**, 65.
[103] W. K. Kappel and R. E. Olson, *Arch. Biochem. Biophys.*, 1984, **235**, 521.
[104] O. Manousek and P. Zuman, *Collect. Czech. Chem. Commun.*, 1964, **29**, 1432.

[105] J. Llor and M. Cortijo, *J. Chem. Soc. Perkin Trans.*, 1977, **2**, 1715.
[106] J. Llor, E. Lopez-Cantarero and M. Cortijo, *Bioelectrochem. Bioenerg.*, 1978, **5**, 276.
[107] R. Izquierdo, M. Blazquez, M. Dominguez and F. Garcia-Blanco, *Bioelectrochem. Bioenerg.*, 1984, **12**, 25.
[108] R. Izquierdo, M. Dominguez, F. Garcia-Blanco and M. Blazquez, *J. Electroanal. Chem.*, 1989, **266**, 357.
[109] J. Volke, *Z. Phys. Chem., Sonderheft*, 1958, 268.
[110] O. Manousek and P. Zuman, *J. Electroanal. Chem.*, 1959–60, **1**, 324.
[111] O. Manousek and P. Zuman, *Collect. Czech. Chem. Commun.*, 1962, **27**, 486.
[112] E. Jacobsen and T. M. Tommelstad, *Anal. Chim. Acta*, 1984, **162**, 379.
[113] J. P. Hart and P. J. Hayler, *Anal. Proc.*, 1986, **23**, 439.
[114] P. Soderhjelm and J. Lindquist, *Analyst*, 1975, **100**, 349.
[115] J. Ballentine and A. D. Woolfson, *J. Pharm. Pharmacol.*, 1980, **32**, 353.
[116] E. Wang and W. Hou, *Microchem. J.*, 1988, **37**, 338.
[117] S. Allenmark, E. Hjelm and U. Larsson-Cohn, *J. Chromatogr.*, 1978, **146**, 485.
[118] B. Lequeu, J. C. Guilland and J. Klepping, *Anal. Biochem.*, 1985, **149**, 296.
[119] S. Kwee, Bioelectrochem. Bioenerg., 1983, **11**, 467.
[120] E. Jacobsen and M. W., Bjornsen, *Anal. Chim. Acta*, 1978, **96**, 345.
[121] L. Rozanski, *Analyst*, 1978, **103**, 950.
[122] J. M. Fernandez Alvarez, A. Costa Garcia, A. J. Miranda Ordieres and P. Tunon Blanco, *J. Electroanal. Chem.*, 1987, **225**, 241.
[123] J. M. Fernandez Alvarez, A. Costa Garcia, A. J. Miranda Ordieres and P. Tunon Blanco, *J. Pharm. Biomed. Anal.*, 1988, **6**, 743.
[124] D. Luo, *Anal. Chim. Acta*, 1986, **189**, 277.
[125] N. Abo El Maali, J. C., Vire, G. J. Partriarche and M. A. Ghandour, *Analysis*, 1989, **17**, 213.
[126] J. Ballentine and A. D. Woolfson, *J. Pharm. Pharmacol.*, 1980, **32**, 353.
[127] C. M. Baugh and C. L. Krumdiek, *Ann. N. Y. Acad. Sci.*, 1971, **186**, 7.
[128] J. M. Scott and D. G. Weir, *Clin. Haematol.*, 1976, **5**, 138.
[129] V. Herbert, A. R. Larrabee and I. M. Buchanan, *J. Clin. Invest.*, 1962, **41**, 1134.
[130] D. L. Longo and V. Herbert, *J. Lab. Clin. Med.*, 1976, **87**, 138.
[131] J. Lankelma, E. van der Kleijn and M. J. Th. Jansen, *J. Chromatogr. Biomed. Appl.*, 1980, **182**, 35.
[132] M. Kohashi, K. Inoue, H. Sotobayashi and K. Iwai, *J. Chromatogr. Biomed. Appl.*, 1986, **382**, 303.
[133] E. Jacobsen and K. B. Thorgersen, *Anal. Chim. Acta*, 1974, **71**, 185.
[134] J. M. Moore, *J. Pharm. Sci.*, 1969, **58**, 1117.
[135] K. Kamata, T. Hagiwara, M. Takahashi, S. Uehara, K. Nakayamaa and K. Akiyama, *J. Chromatogr.*, 1986, **356**, 326.
[136] M. Youssefi and R. L. Birke, *Anal. Chem.*, 1977, **49**, 1380.
[137] H. Sawamoto, *J. Electroanal. Chem.*, 1985, **195**, 395.
[138] L. Calvo, J. Rodriquez, F. Vinagre and A. Sanchez, *Analyst*, 1988, **113**, 321.
[139] S. P. Perone and W. J. Kretlow, *Anal. Chem.*, 1966, **38**, 1760.
[140] L. A. Pachla and P. T. Kissinger, in *Methods in Enzymology*, Vol. 62, D. B. McCormick and L. D. Wright (Eds), Academic Press, 1979, p 15.
[141] G. Sontag and G. Kainz, *Mikrochim. Acta*, 1978, **1**, 175.
[142] D. Amin, *Microchem. J.*, 1983, **28**, 174.
[143] O. Wah Lau, *J. Sci. Food Agri.*, 1985, **36**, 733.
[144] S. Kozar, A. Bujak, J. E. Trifunovic and G. Kniewald, *Z. Anal. Chem.*, 1988, **329**, 760.
[145] M. J. Gomes Silverio, *Rev. Port, Quim.*, 1965, **3**, 154.
[146] R. S. Owen and W. F. Smyth, *J. Food Technol.*, 1975, **10**, 263.
[147] M. Sabino and M. A. P. Zorzetto, *Rev. Inst. Adolfo Lutz*, 1987, **44**, 101.
[148] P. Cherdkiatgumchai and D. R. Grant, *Cereal Chem.*, 1987, **64**, 288.
[149] P. Branca, *Boll. Chim. Union Ital. Lab. Prov., Part Sci.*, 1980, **6**, 143.
[150] U. Gerhardt and R. Windmueller, *Fleischwirtschaft*, 1981, **61**, 1389.
[151] J. Lindquist, *Analyst*, 1975, **100**, 339.
[152] J. Ballantine and A. D. Woolfson, *J. Pharm. Pharmacol.*, 1980, **32**, 353.
[153] W. A. Mason, T. D. Gardner and J. T. Stewart, *J. Pharm. Sci.*, 1972, **61**, 1301.
[154] S. A. Wring and J. P. Hart, *Anal. Chim. Acta*, 1990, **229**, 63.
[155] M. Petersson, *Anal. Chim. Acta*, 1986, **187**, 333.
[156] J. F. Evans, T. Kuwana, M. T. Henne and G,. P. Royer, *J. Electroanal Chem.*, 1977, **80**, 406.
[157] M. F. Dautartas and J. F. Evans, *J. Electroanal Chem.*, 1980, **109**, 301.
[158] K. N. Kuo and W. Murray, *J. Electroanal Chem.*, 1982, **131**, 37.
[159] J. Facci and W. Murray, *J. Electroanal Chem.*, 1982, **54**, 772.
[160] K. Matsumoto, K. Yamada and Y. Osajima, *Anal. Chem.*, 1981, **53**, 1974.

[161] A. G. Ewing, M. A. Dayton and R. M. Wightman, *Anal. Chem.*, 1981, **53**, 1842.
[162] M. P. Brazell and C. A. Marsden, *Br. J. Pharmacol.*, 1982, **75**, 539.
[163] D. T. Fagan, I. Hu and T. Kuwana, *Anal. Chem.*, 1985, **57**, 2759.
[164] M. R. Deakin, P. M. Kovach, K. J. Stutts and R. M. Wightman. *Anal. Chem.*, 1986, **58,** 1474.
[165] K. Iryama, M. Yoshiura and T. Iwamoto, *J. Liq. Chromatogr.*, 1985, **8**, 333.
[166] K. Iryama, M. Yoshiura, T. Iwamoto and Y. Ozaki, *Anal. Biochem.*, 1984, **141**, 238.
[167] W. A. Behrens and R. Madere, *Anal. Biochem.*, 1987, **165**, 102.
[168] W. Lee, P. Hamernyik, M. Hutchinson, V. A. Raisys and R. F. Labbe, *Clin. Chem.*, 1982, **28**, 2165.
[169] D. J. Green, and R. L. Perlman, *Clin. Chem.*, 1980, **26**, 796.
[170] C. S. Tsao and S. L. Salimi, *J. Chromatogr.*, 1982, **245**, 355.
[171] C. S. Tsao, M. Young and S. M. Rose, *J. Chromatogr.*, 1984, **308**, 306.
[172] M. A. Kutnink, J. H. Skala, H. E. Sauberlich and S. T. Omaye, *J. Liq. Chromatogr.*, 1985, **8**, 31.
[173] L. A. Pachla and P. T. Kissinger, *Anal. Chem.*, 1976, **48**, 364.
[174] L A. Pachla and P. T. Kissinger, in *Methods in Enzymology, Vitamins and Coenzymes*, Vol 62, D. B. McCormick and L. D. Wright (Eds), Academic Press, New York, 1979, p 154.
[175] K. V. Thrivikraman, C. Refshauge and R. N. Adams, *Life Sci.*, 1974, **15**, 1335.
[176] R. S. Carr and J. M. Neff, *Anal. Chem.*, 1980, **52**, 2428.
[177] S. A. Margolis and P. T. Davis, *Clin. Chem.*, 1988, **34**, 2217.
[178] K. Ravichandran and R. P. Baldwin, *J. Liq. Chromatogr.*, 1984, **7**, 2031.
[179] J. Wang, and B. Freiha, *Anal. Chem.*, 1984, **56**, 2266.
[180] L. A. Pachla, D. L. Reynolds and P. T. Kissinger, *J. Assoc. Off. Anal. Chem.*, 1985, **68**, 1.

5

Selected coenzymes

5.1 INTRODUCTION

In Chapter 4 the application of electroanalytical techniques to the determination of vitamins was described; the coenzyme forms of some of these were included where the method had been designed to measure both species. In this chapter, coenzymes that it was not convenient to describe in the previous chapters are discussed.

5.2 ELECTROANALYSIS OF SELECTED NUCLEOTIDES WITH COENZYME FUNCTIONS

5.2.1 Nicotinamide adenine dinucleotides

5.2.1.1 Methods involving conventional electrodes

Nicotinamide adenine dinucleotide exists in the oxidized and reduced forms, which are commonly denoted as NAD^+ and NADH respectively (the nomenclature of nucleotides was described in Chapter 2). These compounds function as coenzymes in many biological oxidation/reduction processes so it is not surprising that electrochemical methods have been developed for their determination. The electrochemical oxidation of NADH at rotating glassy carbon and platinum electrodes was studied by Blaedel and Jenkins [1]; it was found that the magnitude of the shift in $E_{1/2}$ with pH did not agree with that predicted for the reaction

$$NADH \rightleftharpoons NAD^+ + 2e^- + H^+ \tag{5.1}$$

From this observation and other studies it was concluded that the electrode surface was chemically involved in the oxidation process and did not merely act as an inert sink for electrons. In the same study, [1] the authors utilized direct oxidation at a rotating glassy carbon electrode for the determination of the reduced form of the coenzyme. The method involved amperometry with a supporting electrolyte containing 0.1 M NaCl–0.005 M phosphate, pH 7.8. The electrode was conditioned before use at +1.3 V for 2 min and −1.3 V for 2 min; pretreatment usually consisted of two such cycles. Using the conditions mentioned, and an electrode rotation speed of 780 rpm, the current–potential curve exhibited a plateau around +0.65 V; it was found that the method could be used to determine NADH amperometrically at the 10-μM level with standard deviations around 1%. Similar studies on NADH have been reported by other workers [2–4] and similar sensitivities were found using amperometric methods. A slightly lower detection limit (5 μM) was reported by Moiroux and Elving [5] using cyclic voltammetry; the method involved electrochemical treatment to produce an adsorbed layer of NAD^+ which eliminates interference from an adsorption pre-wave. Further investigations into the occurrence of such adsorption behaviour have also been reported by Carelli *et al.* [6] and Yakamura *et al.* [7].

Webber *et al.* [8] have exploited the adsorption characteristics of acid-hydrated NADH (I), in a very sensitive voltammetric method for the determination of

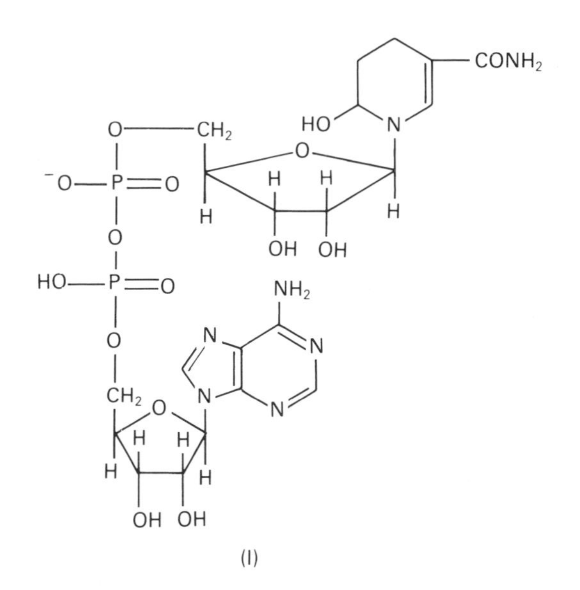

(I)

NADH. In this method, NADH was first dissolved in 10 cm^3 of pH 2.7 solution (succinic acid); after 15 min when the NADH had been converted to (I), 2 cm^3 of 0.1-M NaOH was added to bring the pH to 6.1. Adjusting the pH to 6.1 prevents further hydration reactions from occurring. Square-wave CSV of (I) was then carried out at a static mercury drop electrode (operated as an HMDE) in a buffer at pH 3.6; an accumulated time of 10 s (or 50 s for high sensitivity) and an initial potential of −0.55 V were employed. The calibration plot showed that the peak current levelled off completely and became independent of concentration above 20 μM. This is consistent with the process being due to the reduction of an adsorbed species that at a concentration of 20 μM completely covers the electrode surface. Fig. 5.1 shows the

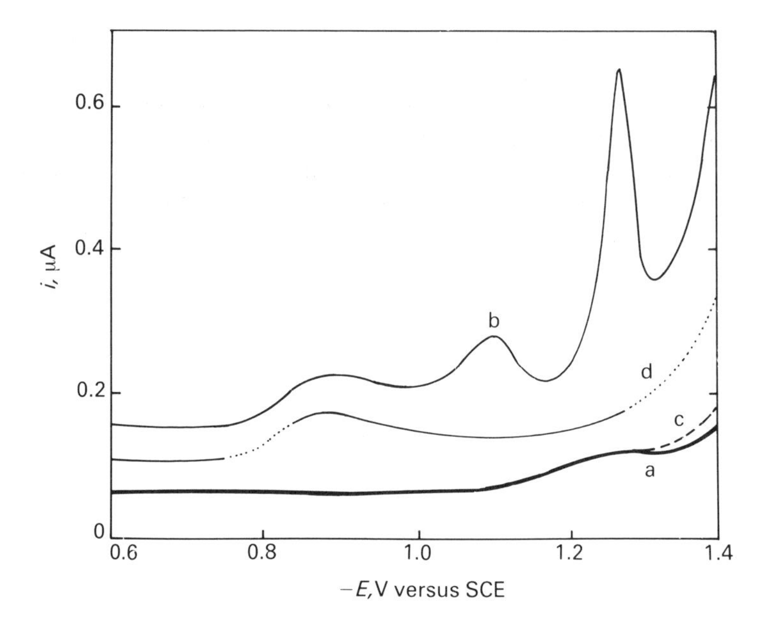

Fig. 5.1 — Square-wave voltammograms of 1.05-μM pretreated NADH in 1.5-M ionic strength succinate buffer with and without 0.0007% Triton X-100 present, (a) Background voltammogram; (b) with NADH added; (c) background with Triton X-100 added; (d) with both NADH and Triton X-100 added. Initial potential −0.55 V; final potential −1.4 V; step height 5 mV; frequency 30 Hz; square-wave amplitude 25 mV; delay time 10 s. (Reproduced from [8] by permission of the copyright holders, Elsevier Science Publishers Physical Sciences and Engineering Division.)

effect of adding Triton X-100 to the sample; it appears that the strongly adsorbing surfactant displaces the acid-hydrated form of NADH from the electrode surface. It was suggested that (I) was properly adsorbed via the adenine ring and that the cathodic peak was the result of a four-electron reduction process involving the protonated adenine ring [9]. The detection limit was less than 7 nM, and the range of linear response covered 2–3 orders of magnitude of NADH concentration. The authors considered that the technique could be used to analyse biological samples if a chromatogaphic step was incorporated into the method.

5.2.1.2 Methods involving chemically modified electrodes and mediated reactions

In order to reduce the overvoltage required for NADH oxidation, electrodes have been modified with a variety of chemical agents. The resulting lower detection potentials are more analytically useful then those required with bare electrodes because better selectivity and sensitivity are obtained; some examples involving mediators for the electrochemical detection of nicotinamide adenine coenzymes are described in this subsection.

Jaegfeldt *et al.* [10] investigated the catalytic oxidation of reduced nicotinamide adenine dinucleotide by graphite electrodes modified with adsorbed aromatics containing catechol functionalities. It was found that 4-[2-(2-naphthyl)vinyl]catechol

($NSCH_2$; II) could be adsorbed directly onto graphite electrodes via the naphthalene ring.

To achieve this a solution was prepared containing the catalyst dissolved in methanol–ethanol 80 : 20; to this was added an equal volume of aqueous 0.1-M hydrochloric acid and the resulting solution contained 0.5 mM of the catalyst. The graphite rods were dipped into this solution under slow agitation for 3–10 min depending on the desired coverage. The catalytic oxidation of NADH mediated by

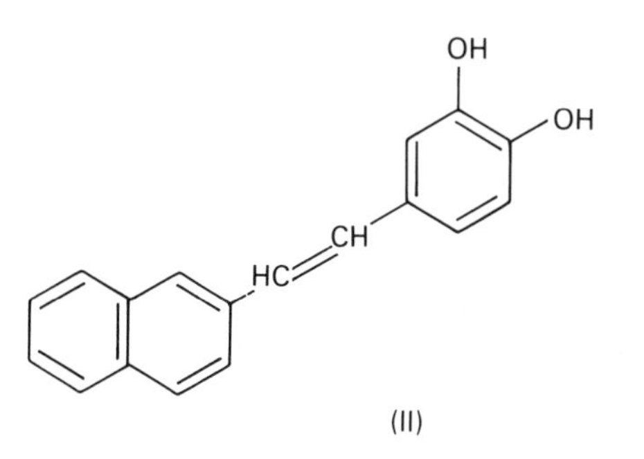

adsorbed $NSCH_2$ is clearly seen from the cyclic voltammograms shown in Fig. 5.2;

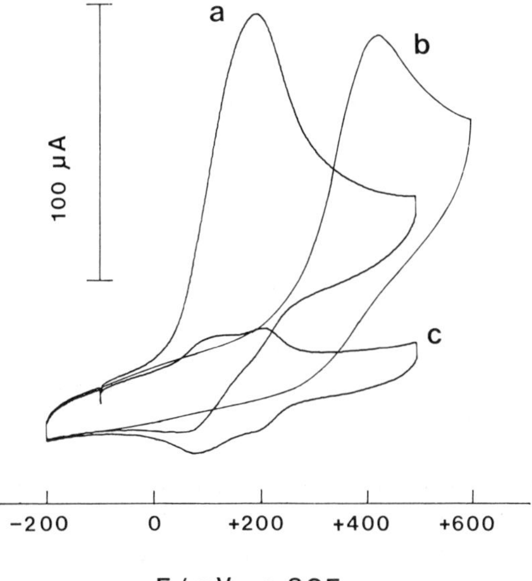

Fig. 5.2 — Catalytic oxidation of NADH mediated by adsorbed $NSCH_2$ (curve a), 2.3-mM NADH, pH 7.0, sweep rate 50 mV s^{-1}. Curve b shows oxidation at a naked graphite electrode in the same solution. Curve c shows $NSCH_2$-modified electrode in a buffer without NADH. Starting potential was −0.10 V versus SCE. (Reprinted with permission from H. Jaegfeldt, A. B. C. Torstensson, L. G. O. Gorton and G. Johansson, *Anal. Chem.*, 1981, **53**, 1979, Copyright 1981, American Chemical Society.)

the overvoltage decreased from 410 mV versus SCE at the unmodified electrode to 185 mV at the chemically modified electrode at pH 7.0. The overall reaction may be described by Eq. (5.2)–(5.4).

$$NSC + NADH + H^+ \xrightarrow{k_1} NSCH_2 + NAD^+ \quad (5.2)$$

$$NSCH_2 \rightleftharpoons NSC + 2H^+ + 2e^- \quad (5.3)$$

$$NADH \rightleftharpoons NAD^+ + H^+ + 2e^- \quad (5.4)$$

It was suggested that the rate-determining step was that shown in Eq. (5.2) with a rate constant k_1; the rate of the electrochemical rection shown in Eq. (5.3) was fast compared with that of the chemical reaction of Eq. (5.2). It should be noted that Eq. (5.3) corresponds to the oxidation of the catechol functional groups to produce the corresponding quinone; as seen in Fig. 5.2b this occurs in two one-electron steps. The overall mediated oxidation reaction is given in Eq. (5.4). It was found that the peak current was proportional to the NADH concentration in the range 0.7 to 7 mM.

In more recent studies, Gorton and colleagues [11,12] have investigated the 7-dimethylamino-1,2-benzophenoxazinium ion (Meldola blue, MB^+) for the catalytic oxidation of the reduced nicotinamide coenzymes. It was shown that the resulting CME could be operated as a sensor for NADH at 0 V versus SCE; this might be expected to further enhance the selectivity of the technique. For analytical purposes, these workers [11,12] used a flow-injection system and the electrochemical cell was of the wall-jet type. The working CME was prepared by dipping pre-treated graphite rods into a 0.1-M cold solution of Meldola blue in a 40% ethanol/1% triethylamine/59% 0.1 M phosphate buffer, pH 3.5. The proposed reaction scheme for the Meldola blue-mediated reaction was considered to be [11].

$$NADH + MB^+ \rightleftharpoons NADH{\cdot}MB^+ \quad (5.5)$$

$$NADH{\cdot}MB^+ \rightarrow NAD^+ + MBH \quad (5.6)$$

$$MBH \rightleftharpoons MB^+ 2e^- + H^+ \quad (5.7)$$

$$NADH \rightarrow NAD^+ + 2e^- + H^+ \quad (5.8)$$

It was found that samples of NADH, measured by the flow-injection system, produced linear calibration plots from 1 μM to 10 mM with an intercept of typically 0.2 μM; the reproducibility was reported to be 0.2–0.6% RSD ($n = 6$) at each concentration. The sensitivity was slightly better at pH 6.0 than at pH 7.0. NADPH was found to give a lower response than NADH, which was considered to be a consequence of the larger diffusion coefficient in the case of the former. Gorton and colleagues [12] have applied the Meldola blue electrode to the determination of glucose using a flow system; it was shown that the method could be used to determine down to 0.25 μM for a 50-μl sample.

A further example of this type was reported by Schelter-Graph *et al.* [13]. In this case the working electrode was fabricated by impregnating a graphite rod with the catalyst 3-β-naphthoyl-Nile Blue [14]; the system, was used to determine lactate in butter, glutamate in beef cubes and ethanol in beer, Wang and Golden [15] very recently showd that a metalloporphyrin manganese (II) *meso*-tetraphenylporphine

(TPP) acetate) chemically modified glassy carbon electrode could be used as a voltammetric sensor; the overpotential for NADH was reduced by 127 mV compared to bare glassy carbon at a pH of 7.0.

The determination of reduced nicotinamide adenine dinucleotide by flow injection analysis using an immobilized-enzyme amperometric system was reported by Chow *et al.* [16]. In this flow injection system the proton acceptor, Bind-Schedlers Green, was reduced by dihydrolipoamide reductase immobilized on glass beads in the presence of NADH. The reduced form of the mediator was then detected amperometrically at +0.04 V versus Ag/AgCl. The carrier solution was 0.1 M phosphate buffer (pH 7.0–8.5) containing 10 μM of mediator and the detection limit was found to be 1 pmol NADH. In the determination of NADH in biological fluids, bilirubin, haemoglobin and ascorbic acid did not interfere.

A slightly different approach was recently described by Yao *et al.* [17] which involved a liquid chromatography system for the specific and simultaneous detection of nicotinamide coenzymes; this was constructed by combining an immobilized glucose-6-phosphate dehydrogenase reactor with an amperometric system based on a phenazine methosulphate-mediated reaction, after separation on a reversed-phase column (Fig. 5.3). The function of the enzyme reactor was to generate NADH or

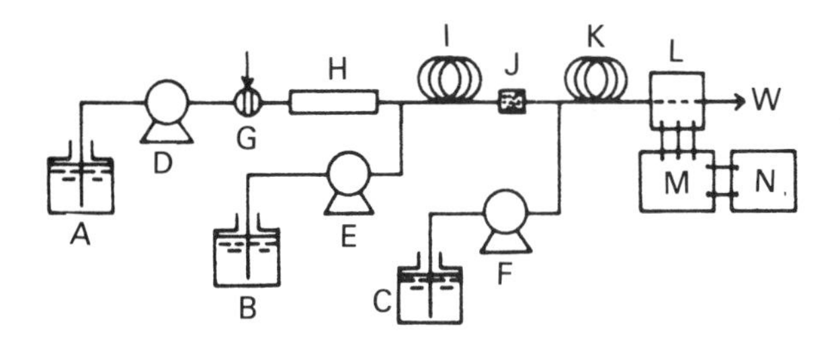

Fig. 5.3 — The liquid chromatography system for amperometric detection of nicotinamide coenzymes with immobilized enzyme: (A) mobile phase (0.02-M phosphate buffer, pH 7.0); (B) reagent I (0.2-M phosphate buffer pH 8.5, containing 3-mM glucose-6-phosphate); (C) reagent II (0.1-M phosphate buffer, pH 6.0, containing 0.4-mM phenazine methosulphate; (D) L. C. pump, 1.0 $cm^3 min^{-1}$; (E) and (F) pumps, 0.5 $cm^3 min^{-1}$; (G) injector; (H) ODS-A reversed-phase column (25 cm × 4.6 mm i.d.) kept constant at 40 ± 0.2°C with a constant-temperature bath; (I) mixing coil (1.5 m × 0.5 mm i.d.); (J) immobilized glucose-6-phosphate dehydrogenase reactor (5 mm × 4 mm i.d.); (K) reaction coil (1.5 m × 0.5 mm i.d.); (L) flow-through platinum electrode; (M) potentiostat; (N) recorder; (W) waste. (Reproduced from [17] by permission of the copyright holders, Elsevier Science Publishers Physical Sciences and Engineering Division).

NADPH by the following reaction:

$$\text{Glucose-6-phosphate} + NAD^+ (NADP^+) \xrightarrow{G6PD} \text{Gluconolactone-6-phosphate} + \text{NADH (NADPH)} \qquad (5.9)$$

In the reaction coil, endogeneous and enzymatically produced NADH and NADPH react with PMS^+ in reagent II to produce PMSH; this is monitored amperometrically at a potential of 0.0 V versus Ag/AgCl according to equations 5.9 and 5.10:

$$NADH(NADPH) + PMS^+ \rightarrow NAD^+ \ (NADP^+) + PMSH \tag{5.10}$$

$$PMSH \xrightarrow[\text{(0.0 V versus Ag/AgCl)}]{\text{Pt anode}} PMS^+ + H^+ + 2e^- \tag{5.11}$$

Fig. 5.4 shows a chromatogram for a solution containing 1 nmol each of the four

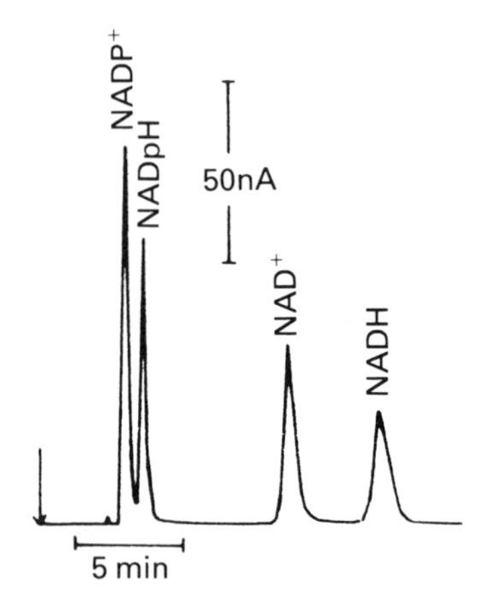

Fig. 5.4 — Typical chromatogram of a standard solution containing 1 mol each of four nicotinamide coenzymes. (Reproduced from [17] by permission of the copyright holders, Elsevier Science Publishers Physical Sciences and Engineering Division.)

coenzymes and elution was complete within 17 min. The calibration graphs were found to be linear from 0.05 to 20 nmol/10-μl injection for all four enzymes. The limits of detection were 3.2, 5.2, 7.9 and 9.4 pmol for $NADP^+$, NADPH, NAD^+ and NADH respectively. The relative standard deviations on 0.5 nmol was found to be 1.2–3.0% ($n = 5$).

5.2.2 Other nucleotides

This short section describes the application of various electrochemical techniques to the determination of a few important nucleotides that act as coenzymes (see Table 5.1).

5.3 ELECTROANALYSIS OF PTERINS: METHODS INVOLVING LCEC

The coenzyme group known as pterins are of interest because of their possible role in several diseases [25]; abnormal levels of various pterins have been observed in several diseases such as phenylketonuria, rheumatoid arthritis, Parkinson's disease and kidney dysfunction. The pterins are a family of heterocyclic compounds that

Table 5.1 — Electroanalytical methods for the determination of flavin nucleotides and coenzyme A

Coenzyme	Electrochemical technique	Supporting electrolyte/ mobile phase	Comments	Ref.
Flavin adenine mononucleotide (FMN) and flavin adenine dinucleotide (FAD)	Thin-layer voltammetry		Study on electrical behaviour of coenzymes. Reduction process involved semiquinone formation at moderate pH values.	[18]
FAD	Cyclic voltammetry at graphite, glassy carbon, platinum and gold electrodes	pH 1.05, KCl/HCl pH 3.0–8.0, citric acid-Na_2HPO_4 pH 2, 9, 10, 11 Merck	Study on adsorption characteristics at electrode surfaces. Electrodes immersed in pH-4.0 buffer with 1-mM FAD, followed by rinsing and medium exchange to plain buffer. Greatest adsorption was found on the graphite electrodes; film coverage was 2×10^{-8} mol cm^2 after 4 h.	[19]
FMN	Adsorptive stripping voltammetry at an HMDE	0.001-M NaOH	Limit of detection was 5×10^{-10} M using 5 min accumulation time and −0.2-V accumulation potential. Stripping peak gave $E_p = -0.54$ V versus Ag/AgCl. Precision was 2% at 4×10^{-8} M ($n = 8$).	[20]
FAD	LCEC using amperometric detection with vitreous carbon electrode; applied potential −0.28 V versus Ag/AgCl. Column, Hypersil APS NH_2 (5 μm).	Acetonitrile–0.1 M-NaH_2PO_4 (3:7), pH 3.05	Calibration graphs rectilinear 2–500 pmol, detection limit 1.25 pmol. Recovery of 2-nmol FAD added to fish liver was above 95% precision, 3.0%.	[21]
CoASH	d.c. polarography/amperometry	Tris-HCl buffer at pH 8.0 (20 μM) + 10 mM $MgCl_2$ + 1-mM EDTA	Measurement of CoASH at −0.2 V versus SCE to determine enzyme activity.	[22–24]

includes pterin itself (III), biopterin (IV), xanthopterin (V) and neopterin (VI); these structures are shown in the fully oxidized states but they also occur naturally in the partially reduced and fully reduced states [26].

The electrochemical characteristics of the pterins have been reported [25,27]. The fully oxidized species undergo a reversible reduction to 5,8-dihydropterins, which rapidly tautomerize to 7,8-dihydropterins. These forms can undergo further reduction to produce 5,6,7,8-tetrahydropterins, or are oxidized back to the fully oxidized state. The tetrahydropterins are reversibly oxidized to quinonoid dihydropterins, which can again undergo tautomerism to 7,8-dihydropterins. This reaction pathway is shown in Fig. 5.5. An exception to this scheme occurs with xanthopterin (VI), in which an enolic hydroxyl group in the 6-position blocks reduction to the terahydro form.

The electrochemical behaviour has been exploited for the sensitive measurement of a variety of pterins, and their different oxidation states, in biological samples.

On one of the first reports by Kissinger's group [28] on LCEC methods for pterin analysis, an amperometric detector equipped with a single glassy carbon electrode, operated in the reductive mode, was used. Samples of urine were initially treated to obtain pterins in the oxidized forms; after centrifugation the pterins were separated from the supernatant by ion exchange chromatography. Analysis was carried out with a reversed-phase column and a mobile phase containing an ion-pairing agent (octyl sulphate), and the electrode was operated at a potential of −0.7 V versus Ag/AgCl. The major pterins in urine were reported to be neopterin and biopterin while xanthopterin, 6-hydroxymethylpterin, pterin and pterin-6-carboxylic acid are found in much lower concentrations. The electrochemical reduction potentials of these compounds are shown in Fig. 5.6. The same method was applied to the determination of pterins in mouse brains and liver samples. The high sensitivity of the method may be seen from the detection limits shown in Table 5.2.

Further work has been carried out by Kissinger's group [26] in order to measure simultaneously various naturally occurring pterins in their different oxidation states. These workers investigated LCEC with dual glassy carbon electrodes in the parallel configuration; one electrode was set at −0.700 V to measure fully oxidized forms of pterins, and the other was held at +0.800 V to measure dihydro and tetrahydro species. Fig. 5.7 shows the chromatogram obtained from a sample of human urine; this demonstrates clearly the ability of this technique to monitor the three possible oxidation states of the pterins simultaneously. Peak identification for chromatograms of urine was confirmed by poising the two electrodes at different positions on the voltammetric wave and then measuring the current ratios; for example, for the tetrahydropterins the electrode potentials were set at + 0.30 V and +0.50 V. The peak current ratios for tetrahydroneopterin and tetrahydrobiopterin were 0.59 and 0.63 in urine and in standards the same compounds gave ratios of 0.61 and 0.61 respectively. Therefore, good agreement was obtained between the two sets of results, which confirmed the identity of the peaks.

The same research group [29] have continued their studies on pterins and have investigated the electrochemistry of 6-methylpterin using a dual-electrode amperometric detector in the series configuration; this was used to determine coupled electrochemical reactions. The parallel arrangement was also used to study the rate

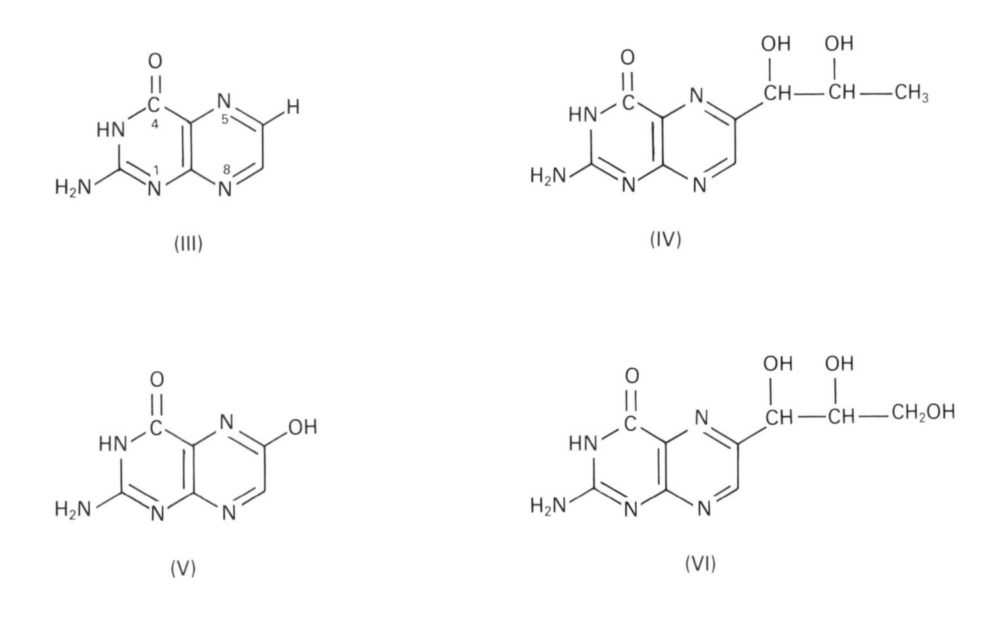

of tautomerism of quinonoid 6-methyldihydropterin. A further LCEC investigation was carried out which involved the use of 6-methyl-5,6,7,8-tetrahydropterin to determine phenylalanine hydroxylase activity [30]; this was achieved by monitoring the rate of oxidation to the dihydro species. It was also possible to monitor the enzymatic reaction by measuring the rate of formation of the dihydro species since this was also electro-oxidizable at the potential employed (+1.2 V versus Ag/AgCl).

The liquid chromatography separation and detection of various pterin species has also been the subject of investigations by other workers. Hyland [31] has developed an interesting method to monitor the three oxidation states using sequential electrochemical and fluorometric detection. Tetrahydropterins were detected at the first electrode of a series dual-electrode coulometric detector that was held at +0.05 V. The dihydropterins were monitored by fluorometric detector following post-column electrochemical oxidation; to achieve this the second electrode was set at +0.4 V and a guard column in series was set at +0.8 V. The fully oxidized pterins were measured by their native fluorescence. The system used in this investigation is shown in Fig. 5.8. It was suggested that the relative selectivity of fluorescence detection and the low potential required to oxidize the tetrahydropterins results in a method which can measure all the oxidation forms in biological samples with minimal sample clean-up. The same author and colleagues [32–34] have used similar methods to monitor naturally occurring pterins in cerebrospinal fluid with a view to studying inborn errors of metabolism in children.

The combination of electrochemical detection, for tetrahydrobiopterin, and electrofluorometric detection for the determination of dihydropterin in plasma was also the approach selected by Powers *et al.* [35].

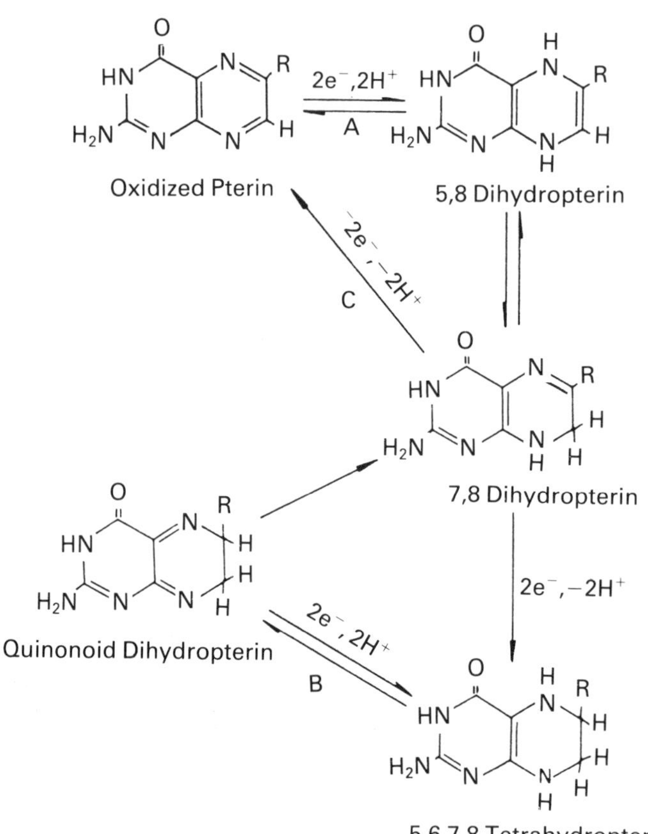

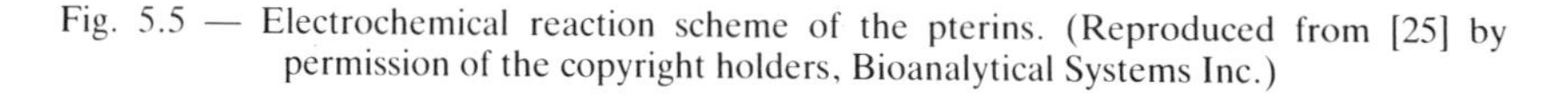
Fig. 5.5 — Electrochemical reaction scheme of the pterins. (Reproduced from [25] by permission of the copyright holders, Bioanalytical Systems Inc.)

5.4 UBIQUINONES/UBIQUINOLS (COENZYME Q)

5.4.1 Methods involving polarographic and voltammetric techniques

The ubiquinones function as electron carries in the mitochondrial electron transport chain; they may also be involved in mediation of oxidative damage to mitochondrial lipids [36]. In addition, ubiquinols have been shown to possess antioxidant properties [37,38].

The naurally occurring forms of this coenzyme group contain the substituted benzoquinone (quinol) moiety with various numbers of isoprene groups attached to the 3-position; the structure of the oxidized forms is shown in (VII).

Where $n = 6$ to 10 in some common ubiquinones: these are usally known as ubiquione-6 to ubiquinone-10.

The mechanism of reduction of ubiquinones has been described by Moret *et al.* [39] and involves the addition of two protons and two electrons to form the corresponding ubiquinols (Fig. 5.9).

Detailed studies on the electrochemical behaviour of ubiquinone-30 in 90%

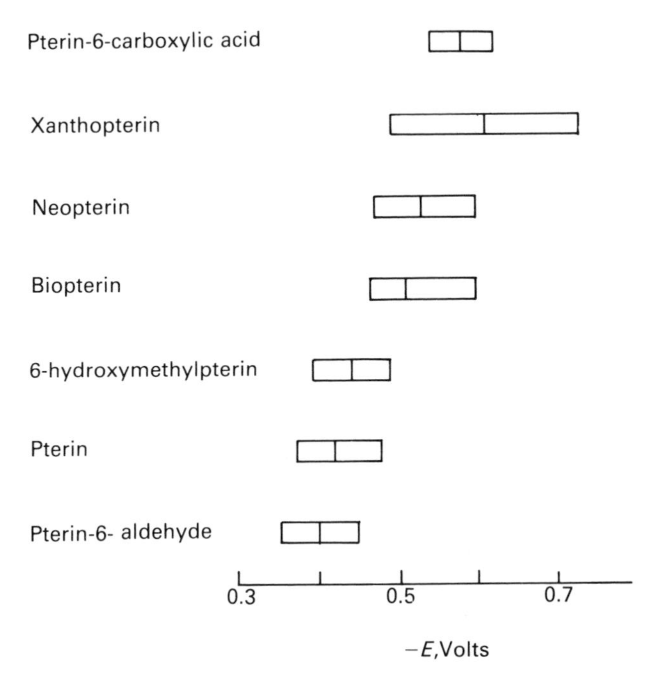

Fig. 5.6 — Diagrammatic presentation of the reduction potentials of selected pterins (vertical lines in each box represent the $E_{1/4}$, $E_{1/2}$ and $E_{3/4}$ potentials). (Reproduced from [28] by permission of the copyright holders, Academic Press Inc.)

Table 5.2 — Detection limits for selected pterins. (Reproduced from [28] by permission of the copyright holders, Academic Press Inc.)

Pterin	LOD (pmol)[a]
Biopterin	0.19
6-Hydroxymethylpterin	0.45
erythro-Neopterin	0.15
threo-neopterin	0.14
Pterin	0.37
Pterin-6-aldehyde	0.20
Pterin-6-carboxylic acid	0.33
Xanthopterin	0.28
Tetrahydrobiopterin	0.75

[a] Limit of detection at a signal-to-noise ratio of 3.

methanolic acetate buffers have been carried by O'Brien and Olver [40]. These authors confirmed that the reduction process involved two protons and two electrons; however, it was shown by cyclic voltammetry that adsorption of reactant and product occurred at the HMDE.

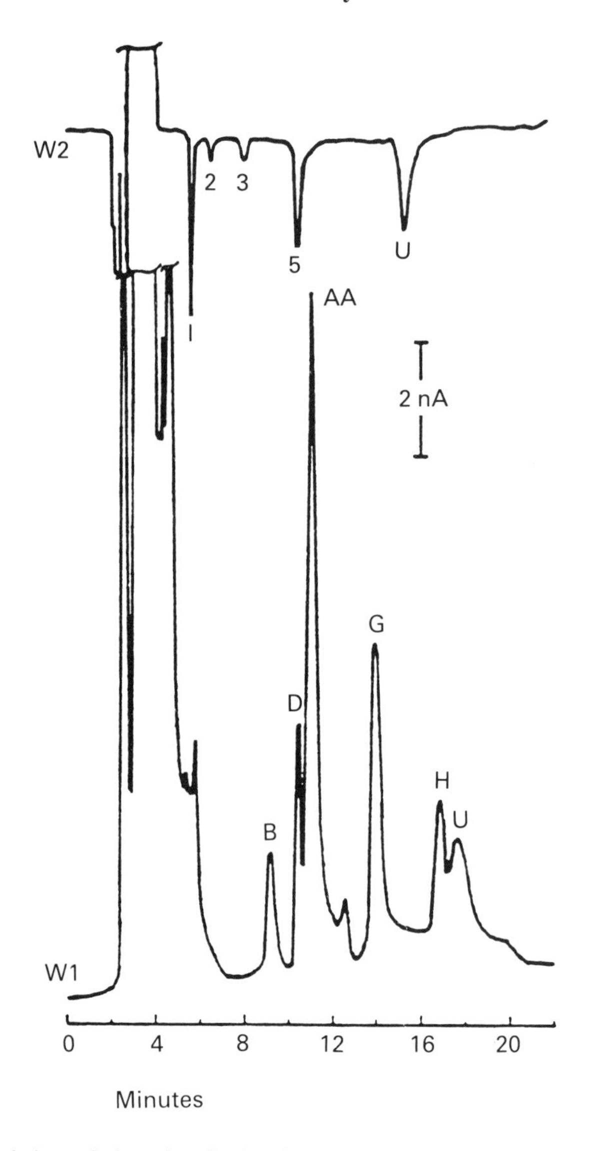

Fig. 5.7 — Dual-electrode detection of reduced and oxidized pterins in urine. Chromatographic conditions: mobile phase, 3-mM octyl sodium sulphate in 0.1-M sodium phosphate buffer, pH 2.5; column, Biophase ODS 5 μm (25 cm × 4.6 mm i.d.); flow rate = 1.0 $cm^3 min^{-1}$. Detector potentials: W1 = +800 mV, W2 = −700 mV. Peak identities: (1) *erythro*-neopterin; (2) *threo*-neopterin; (3) xanthopterin; (5) biopterin. B, tetrahydroneopterin; D, dihydroxanthopterin; G, tetrahydrobiopterin; H, dihydrobiopterin; AA, ascorbic acid; U, unidentified peak. (Reprinted with permission from C. E. Lunte and P. T. Kissinger, *Anal. Chem.*, 1983, **55**, 1458. Copyright 1983, American Chemical Society.)

Erabi *et al.* [41] have applied polarography and linear-sweep voltammetry to studies involving ubiquinone-10; the purpose of the investigation was to study the effect of chromatophores on uniquinone electrochemistry. It was suggested that at

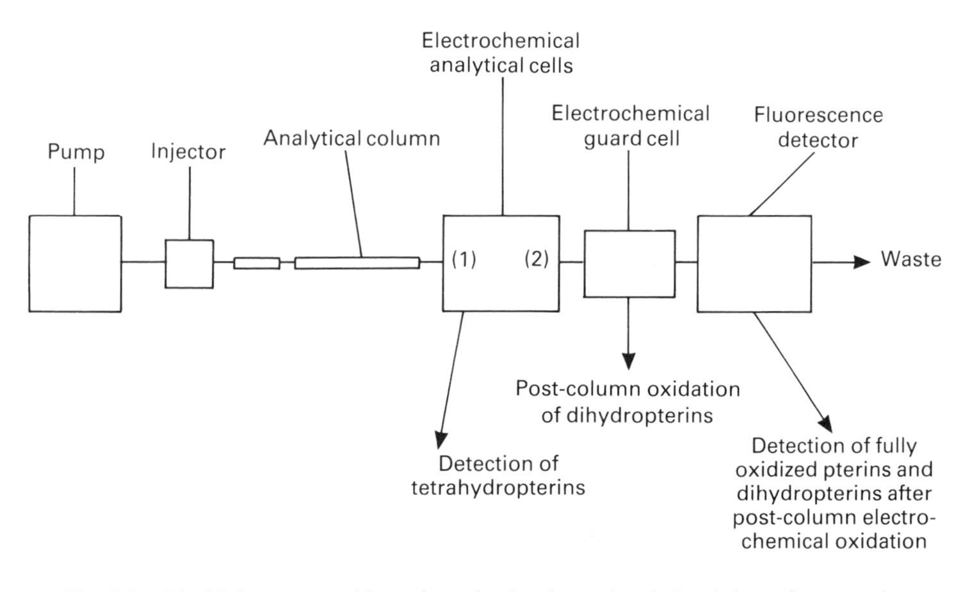

Fig. 5.8 — Liquid chromatographic configuration for electrochemical and electrofluorometric detection. (Reproduced from [31] by permission of the copyright holders. Elsevier Science Publishers Physical Sciences and Engineering Division.)

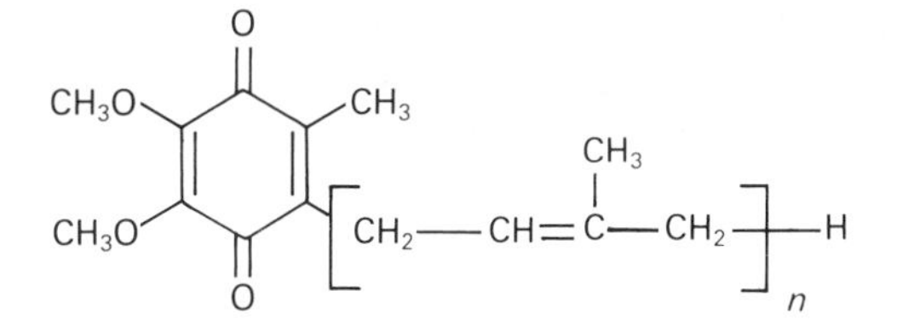

(VII) Ubiquinones

neutral pH the electrochemical reaction of ubiquinone-10 bound with chromatophores initially involved the addition of two electrons and one proton; there was a delay in the addition of the second proton.

Hart and Catterall [42] have performed DPP on solutions containing ubiquinone-10 in 90% ethanol-0.05 M acetate buffer pH 6.0 and showed that a well-defined peak was obtained. In addition, it was shown that extracts from plasma exhibited a DPP peak at −0.3 V versus SCE using the same supporting electrolyte; when standard additions of the coenzyme were made to the extract the peak increased almost

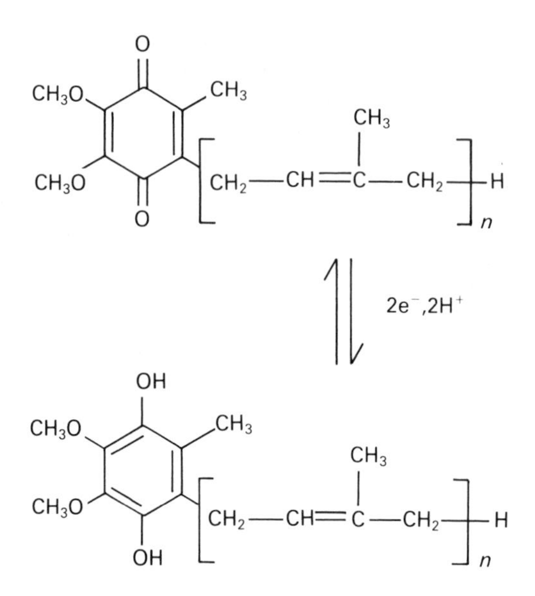

Fig. 5.9 — Mechanism of reduction of ubiquinones.

linearly. Therefore, it was considered likely that the peak was due to ubiquinone-10 and that it may be possible to use DPP for its determination in plasma.

It should be mentioned that since these coenzymes have been shown to exhibit adsorption characteristics at a mercury eletrode [40] it may be possible to develop a method of analysis involving AdSV this might be expected to result in a very sensitive method of analysis. Such methods have been developed for other quinones that exhibited adsorption behaviour [43,44].

5.4.2 Methods involving LCEC

The interest in the determination of ubiquinones and ubiquinols at the low levels found in biological samples has resulted in a variety of liquid chromatographic methods of analysis using amperometric detection systems.

In one of the earliest LCEC methods for the determination of ubiquinones with 8, 9, and 10 isoprene units, Ikenoya *et al.* [45] employed an ODS column with a mobile phase containing $HClO_4$ (70%)–EtOH, (1:999), containing 0.05-M $NaClO_4$. The amperometric detector was equipped with a cell containing a glassy carbon working electrode, which was operated in the reductive mode; this was set at a potential of −0.3 V versus Ag/AgCl. Human serum was extracted with *n*-hexane and the residue, following evaporation of solvent, was dissolved in isopropanol; an aliquot of this was injected onto the reversed-phase column for analysis of ubiquinone-10. The concentrations were found to be in the range 0.54–1.45 $\mu g\,cm^3$ ($n = 5$) and agreed with the results obtained by LCUV; however, the LCEC method was reported to be about 20 times more sensitive than LCUV; the limit of detection for the former was 200 pg.

Further studies by Ikenoya and colleagues [46,47] have involved the simultaneous determination of ubiquinols and ubiquinones in tissues and mitochondria using a combination of electrochemical and UV detection. A reversed-phase system was again employed but in this case the mobile phase consisted of 0.05-M $LiClO_4$ in EtOH–MeOH–70% $HClO_4$, (900:100:1); the retention times of ubiquinol-10 and ubiquinone-10 were 5.7 min and 7.4 min respectively. The reduced form of the coenzyme was determined by LCEC using a detector potential of +0.7 V versus Ag/AgCl and the oxidized species was detected at 275 nm. The authors reported that the method could sensitively and specifically measure the redox state of ubiquinone in mitochondria and tissues.

A similar approach to the simultaneous determination of the reduced and oxidized forms of the coenzymes, as well as α-tocopherol, in biological samples was described by Lang and coworkers [48,49]. Tissue samples were homogenized with water, the homogenate was mixed with the antioxidant 2,6-di-*t*-butyl-*p*-cresol and sodium dodecyl sulphate, and the analytes were extracted with hexane; Fig. 5.10

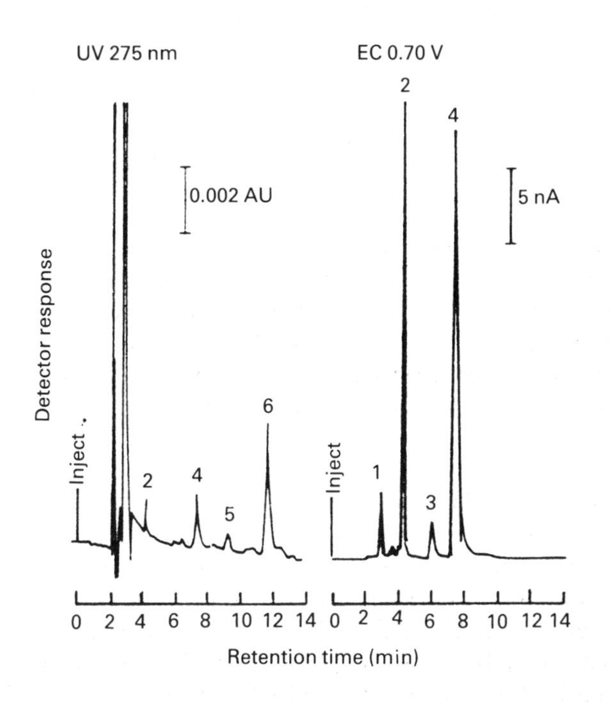

Fig. 5.10 — Ultaviolet and electrochemical chromatograms of guinea pig brown adipose tissue: (1) BHT, (2) α-tocopherol, (3) ubiquinol-9, (4) ubiquinol-10, (5) ubiquinone-9, (6) ubiquinone-10. Mobile phase methanol/reagent alcohol 1:9 (v/v), 20-mM lithium perchlorate. (Reproduced from [48] by permission of the copyright holders, Academic Press Inc.)

shows the resulting chromatograms from a sample of guinea pig tissue. In a recent report, Edlund [50] also used this coupled technique for the determination of ubiquinone-10, α-tocopherol and cholesterol in biological samples.

Another example which demonstrates the potential of reversed-phase LCEC for the determination of trace qualities of ubiquinones was reported by Jang and Soblosky [51]. It was shown that ubiquinones with 6–10 isoprene units could be chemically reduced to the corresponding ubiquinols using sodium borohydride; the limit of detection for ubiquinol-6 was 120 pg using a potential of +0.7 V. It was suggested that the method may be applicable to quantify biological samples in which only small amounts of ubiquinone are present.

This was in fact the technique selected by Okamoto *et al.* [52] for the determination of reduced and total ubiquinones in biological materials. Samples of plasma ($0.2\,cm^3$), urine ($2\,cm^3$) or rat tissue homogenate ($0.5\,cm^3$) were mixed with ethanol and extracted with hexane. The residues were dissolved in ethanol and an aliquot was analysed for ubiquinols; a second aliquot was treated with 0.25% $NaBH_4$ and after 10 min this was analysed for total ubiquinols. The electrochemical detector was operated in the oxidative mode at a potential of +0.6 V versus Ag/AgCl; The calibration graph was linear up to 20 ng of ubiquinol homologues and the coefficient of variation was less than 3.1% for both ubiquinol-10 and ubiquinone-10.

In contrast, Hedrick and White [53] have developed a method involving reductive-mode LCEC for marine bacterial samples containing ubiquinone and menaquinone containing eight isoprene units. The detection system consisted of two vitreous carbon electrodes set at −0.06 and −0.30 V versus SCE; the two compounds gave similar responses at the more positive potential but the menaquinone gave about four times the response at −0.30 V.

5.5 CONCLUSION

This chapter has been concerned with the application of electrochemical techniques to the determination of several structurally different groups of coenzymes.

The few examples described here clearly illustrate the possibilities for the determination of nucleotide coenzymes, pterins and ubiquinones (ubiquinols) in biological materials. For the determination of physiological blood levels, LCEC methods offer both high sensitivity and selectivity; in the case of the pterins it is also possible to monitor several oxidation states in one sample injection by using a cell containing dual parallel electrodes. It has been possible to employ chemically modified electrodes to reduce the overpotential for the oxidation of some nucleotide coenzymes; such electrodes may be employed as detectors in FIA or LCEC and would be expected to give better selectivity than the conventional electrode materials. It may be possible to use similar strategies for the determination of other types of coenzymes. In addition, it also seems feasible that methods for substrates, or enzymes, that are dependent on certain coenzymes may be determined indirectly by this kind of detection system. Therefore, these techniques should prove valuable in the area of biomedicine, particularly in diagnostic testing.

The use of polarographic and voltammetric techniques for the measurement of several coenzymes mentioned earlier has been possible because they are electroactive. It would also appear that some of these undergo adsorption at mercury electrodes and are therefore amenable to stripping voltammetry. Therefore, further work in this area should prove fruitful in devloping sensitive methods that could be of clinical use.

REFERENCES

[1] W. J. Blaedel and R. A. Jenkins, *Anal. Chem.*, 1975, **47**, 1337.
[2] G. G. Guilbault and T. Cserfalvi, *Anal. Lett.*, 1976, **9**, 277.
[3] T. C. Wallace and R. W. Coughlin, *Anal. Biochem.*, 1977, **80**, 133.
[4] T. C. Wallace, M. B. Leh and R. W. Coughlin, *Biotechnol. Bioeng.*, 1977, **19**, 901.
[5] J. Moiroux and P. J. Elving, *Anal. Chem.*, 1979, **51**, 346.
[6] I. Carelli, R. Rosati and A. Casini, *Electrochim. Acta*, 1981, **26**, 1695.
[7] K. Takamura, M. Atsuko and K. Fumiyo, *Bioelectrochem. Bioenerg.*, 1981, **8**, 229.
[8] A. Webber, M. Shah and J. Osteryoung, *Anal. Chim. Acta*, 1984, **157**, 1.
[9] A. Webber and J. Osteryoung, *Anal. Chim. Acta*, 1984, **157**, 17.
[10] H. Jaegfeldt, A. B. C. Torstensson, Lo. G. O. Gorton and G. Johansson, *Anal. Chem.*, 1981, **53**, 1979.
[11] Lo. G. O., Gorton, A. Torstensson, H. Jaegfeldt and G. Johansson, *J. Electroanal. Chem.*, 1984, **161**, 103.
[12] R. Appelqvist, G. Marko-Varga, Lo Gorton, A. Torstensson and G. Johansson, *Anal. Chim. Acta*, 1985, **169**, 237.
[13] A. Schelter-Graph, H. L. Schmidt and H. Huck, *Anal. Chim. Acta.*, 1984, **163**, 299.
[14] H. Huck, *Z. Anal. Chem.*, 1982, **313**, 548.
[15] J. Wang and T. Golden, *Anal. Chim. Acta*, 1989, **217**, 343.
[16] T. Chow, S. Yoshida, M. Itoh, S. Hirose and T. Takeda, *Bunseki Kagaku*, 1984, **33**, 310.
[17] T. Yao, Y. Matsumoto and T. Wasa, *Anal. Chim. Acta*, 1989, **218**, 129.
[18] O. S. Ksenzhek and S. A. Petrova, *Bioelectrochem. Bioenerg.*, 1983, **11**, 105.
[19] Lo. G. O., Gorton and G. Johansson, *J. Electroanal. Chem.*, 1980, **113**, 151.
[20] J. Wang, D. B. Luo, P. A. M. Farias and J. S. Mahmoud, *Anal. Chem.*, 1985, **57**, 158.
[21] C. Cann–Moisan, J. Caroff and E. Girin, *J. Chromatogr.*, 1988, **442**, 441.
[22] P. D. J. Weitzman, in *Electroanalysis in Hygiene, Environmental, Clinical and Pharmaceutical Chemistry*, W. F. Smyth (Ed.), Elsevier, Amsterdam, 1980, p. 137.
[23] P. D. J. Weitzman and H. A. Kinghorn, *FEBS Lett.*, 1978, **88**, 255.
[24] P. D. J. Weitzman, *Biochem. Soc. Trans.*, 1976, **4**, 724.
[25] C. Lunte, *Current Separations*, 1983, **5**, 39.
[26] C. E. Lunte and P. T. Kissinger, *Anal. Chem.*, 1983, **55**, 1458.
[27] L. G. Karber and G. Dryhurst, *J. Electroanal. Chem.*, 1982, **136**, 271.
[28] C. E. Lunte and P. T. Kissinger, *Anal. Biochem.*, 1983, **129**, 377.
[29] C. E. Lunte and P. T. Kissinger, *Anal. Chem.*, 1984, **56**, 658.
[30] C. E. Lunte and P. T. Kissinger, *Anal. Biochem.*, 1984, **139**, 468.
[31] K. Hyland, *J. Chromatogr.*, 1985, **343**, 35.
[32] K. Hyland, I. Smith, D. W. Howells, P. T. Clayton and J. V. Leonard, in *Biochemical and Clinical Aspects of Pteridines*, Vol. 4, H. Wachter, H. C. Curtis, W. Pfleiderer (Eds), Walter de Gruyter and Co., Berlin, New York, 1985.
[33] D. W. Howells and K. Hyland, *Clin. Chim. Acta*, 1987, **167**, 23.
[34] D. W. Howells and K. Hyland, *J. Chromatogr., Biomed. Appl.*, 1986, **54**, 285.
[35] A. G. Powers, J. H. Young and B. E. Clayton, *J. Chromatogr., Biomed. Appl.*, 1988, **76**, 321.
[36] H. Suzuki and T. E. King, *J. Biol. Chem.*, 1983, **258**, 352.
[37] R. Takayanagi, K. Takeshige and S. Minakami, *Biochem. J.*, 1980, **192**, 853.
[38] H. Bindoli, L. Cavallini and P. Jocelyn, *Biochim. Biophys. Acta*, 1982, **681**, 496.
[39] V. Moret, S. Pinamonti and E. Fornasary, *Biochem. Biophys. Acta*, 1961, **54**, 381.
[40] F. L. O'Brien and J. W. Olver, *Anal. Chem.*, 1969, **41**, 1810.
[41] T. Erabi, T. Higuti, T. Kakuno, J. Yamashita, M. Tanaka and T. Horio, *J. Biochem.*, 1975, **78**, 795.
[42] J. P. Hart and A. Catterall, in *Electroanalysis in Hygiene, Environmental, Clinical and Pharmaceutical Chemistry*, W. F. Smyth (Ed.), Elsevier, Amsterdam, 1980, p. 145.
[43] J. C. Vire, V. Lopez, G. J. Patriarche and G. D. Christian, *Anal. Lett.*, 1988, **21**, 2217.
[44] J. C. Vire, N. Abo El Maali and G. J. Patriarche, *Talanta*, 1988, **35**, 997.
[45] S. Ikenoya, K. Abe, T. Tsuda, Y. Yamano, O. Hiroshima, M. Ohmae and K. Kawabe, *Chem. Pharm. Bull.*, 1979, **27**, 1237.
[46] K. Katayama, M. Takada, T. Yazuriha, K. Abe and S. Ikenoya, *Biochem. Biophys. Res. Commun.*, 1980, **95**, 971.
[47] S. Ikenoya, M. Takada, T. Yozuriha, K. Abe, K. Katayama, *Chem. Pharm. Bull.*, 1981, **29**, 158.
[48] J. K. Lang, K. Gohil and L. Packer, *Anal. Biochem.*, 1986, **157**, 106.
[49] J. K. Lang and L. Packer, *J. Chromatogr.*, 1987, **385**, 109.
[50] P. O. Edlund, *J. Chromatogr., Biomed. Appl.*, 1988, **69**, 87.
[51] I. Jeng and J. S. Soblosky, *J. Chromatogr.*, 1984, **295**, 515.
[52] T. Okamoto, Y. Fukunaga, Y. Ida and T. Kishi, *J. Chromatogr.*, 1988, **74**, 11.
[53] D. B. Hedrick and D. C. White, *J. Microbiol. Methods*, 1986, **5**, 243.

Index of chemical substances

General index